Mac Unlocked

Everything You Need to Know
to Get Cracking in macOS Big Sur

David Pogue

Simon & Schuster Paperbacks

NEW YORK   LONDON   TORONTO   SYDNEY   NEW DELHI

Simon & Schuster Paperbacks
An Imprint of Simon & Schuster, Inc.
1230 Avenue of the Americas
New York, NY 10020

Copyright © 2020 by David Pogue

First Simon & Schuster trade paperback edition December 2020

SIMON & SCHUSTER PAPERBACKS and colophon are registered trademarks
of Simon & Schuster, Inc.

For information about special discounts for bulk purchases,
please contact Simon & Schuster Special Sales at 1-866-506-1949
or business@simonandschuster.com.

The Simon & Schuster Speakers Bureau can bring authors to your
live event. For more information or to book an event, contact the
Simon & Schuster Speakers Bureau at 1-866-248-3049
or visit our website at www.simonspeakers.com.

Manufactured in the United States of America

10   9   8   7   6   5   4   3   2   1

Library of Congress Cataloging-in-Publication Data has been applied for.

ISBN 978-1-9821-7667-9
ISBN 978-1-9821-7668-6 (ebook)

ALSO BY DAVID POGUE

Contents

Part Two: Welcome to Big Sur

Part Three: macOS Piece by Piece

Introduction

Apple introduced the first Mac on January 24, 1984. Since that day, tens of thousands of designers, managers, and programmers have come and gone through this massive corporation, leaving their imprints on the Mac and its operating system along the way.

They've added and subtracted jacks and connectors, invented and retired software features, and reconceived software designs over and over again. After 37 years of this, you might expect the Mac operating system to be a hot mess.

Impressively enough, many of the original Mac's founding principles are still at work today. The general philosophy still goes like this: Offer amazing technology, but hide as much of the complexity as possible.

And also this: Give all the different apps roughly the same design, so Mac owners have less to learn. Use the same keyboard shortcuts everywhere. Put the buttons and menus in consistent places.

Oh—and make it all look beautiful. As Steve Jobs said in 2000, upon unveiling Mac OS X: "We made the buttons on the screen look so good you'll want to lick 'em."

A Little History

Every year, Apple introduces a new version of macOS. For many years, these updates were nicknamed for big cats (Jaguar, Tiger, Leopard, and so on). Apple was just about to run out of big cat names—it couldn't quite bring itself to release "Mac OS X Ocelot"—when the policy changed. These days, each release is named after a beloved place in California. There was Mavericks,

Yosemite, El Capitan, Sierra, High Sierra, Mojave, Catalina—and now, at last, macOS Big Sur. (The macOS versions have numbers, too—Big Sur is technically macOS 11—but few people know them.)

And how does Apple entice you to upgrade each year? It really has only one lever to pull: *Add more features.*

Well, also *Make things run faster* and *Periodically redesign things to look cooler.* But mostly it's *Add more features.*

On one hand, adding so much to the Mac means nobody gets left behind. Parents get more internet safety features for their kids. Game-company software engineers get new APIs for driving graphics circuitry. Professionals with mobility limitations get to operate the Mac with head gestures and facial expressions.

On the other hand, pretty soon, there are simply too many nuances and features for the average person to master.

How This Book Was Born

Hello there, I'm David.

I wrote my first Mac book, *Macs for Dummies*, in 1992. It managed to cover every feature of the Mac operating system in 338 pages.

In 2001, I wrote *Mac OS X: The Missing Manual.* The goal, once again, was to cover the entire operating system, all 50 apps that came with it, and all Mac models.

But each year, Apple introduced a new version of the operating system with more features; each year, I updated the book to cover the new stuff; and each year, the book got thicker. Eventually, it was 850 pages long, weighed 3 pounds, and *still* didn't cover everything—I'd started offloading entire chapters as downloadable PDF files.

Through the decades, I've never stopped teaching people the Mac in person, over the internet, or (shudder) over the phone. Those teaching experiences produced a little voice deep inside me, which eventually grew to be a very big voice. And what it kept shouting was, "NOBODY CARES ABOUT EVERYTHING!"

You want to learn the Mac, yes—but good instruction isn't just walking through a thousand features and telling you what each one does. It's also *curation.* It's telling you which features are even *worth* knowing about and explaining when you might use them. It's distinguishing between features Apple is excited to promote (or once was, years ago) and truly great features that never got any marketing love.

That's the idea behind *Mac Unlocked*: to teach you the features you'll actually find valuable, in the right order, with the right emphasis. (As for the title: It's a little pun. Of course, you want to unlock the power within your Mac—but you also have to unlock it with your password or fingerprint before you can even begin.)

Eight Principles of the Mac

So what are those basic elements of the Mac's design that live on after so many years? Here's a sampling—intended not only for historical purposes, but to give you some starter insight into the way macOS works.

- **The menu controls the Mac itself.** At the top left of your screen, at almost all times, sits the menu. It's there no matter what app you're using, and it always contains important commands that affect the entire Mac. They include **System Preferences** (the Mac's settings-control panel), **Restart**, and **Shut Down**.

Menu bar

- **A menu ellipsis (...) means a dialog box will open first.** Some menu commands include an ellipsis (for example, **Restart...**). Those three dots tell you that choosing this command won't do anything immediately; first, a dialog box will appear, in which you have to answer a question or two. (In the case of **Restart...** that question is "Are you sure?")

- **The second menu controls the app you're using.** The menu next to the menu is the *application menu*. It exists for two reasons. First, it tells you what app you're using. It might say **Calendar, Notes,** or **Safari**—or **Finder,** when you're at the desktop.

  Second, it contains commands that pertain to the running of that app—most importantly **Quit**.

- **The blue, lower-right button always means "yes."** Every dialog box offers buttons that let you close the box and proceed. **Cancel** is usually one of them. But the other one—called **OK, Continue, Save, Print,** or whatever—is always blue, always in the lower-right corner, and always the button that means "Yes, let's proceed." It's what Apple calls the *default* button.

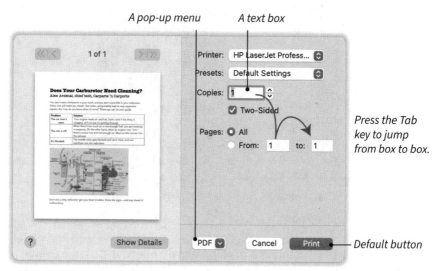

A pop-up menu   *A text box*

Press the Tab key to jump from box to box.

Default button

Dialog boxes

And why do you care? Because you can always "click" the default button by pressing Return on your keyboard, which is usually more efficient than using the mouse or trackpad.

- **You can click any part of things.** It's just a little gift, but a sweet one: You don't have to click squarely on a checkbox or a slider handle. You can be a little sloppier, and the Mac still understands.

Click anywhere on the text—
doesn't have to be the checkbox.

Click anywhere on the slider—
doesn't have to be the handle.

Sloppy clicking

- **The Mac is becoming more like the iPhone and iPad every year.** Little by little, the Mac takes on more of the personality, apps, and design of iOS (the iPhone/iPad software). The apps have the same names, features, and layouts. The features have the same names and are in the same places. Apple has even written an app called Catalyst, intended for software companies, that converts iPad apps into Mac apps. That way, the rest of the world can get in on the great iPaddification of the Mac, too.

Here and there, you can find Mac fans who object. They take Apple's actions to mean the company is losing interest in the Mac, the computer that made it famous. (It doesn't help that the entire Mac business generates less than 10% of Apple's revenue these days.)

But it's also possible to see the Grand Apple Unification as a good thing. First, it means if you've learned one Apple gadget, you've learned them all. Second, it means that keeping the Mac alive is less expensive for Apple, because it's already done so much work on the software for the iPhone and iPad.

- **System Preferences changes everything.** You can get to the Mac's bustling preference-setting center by choosing its name from the menu. Within these 32 or so apps are the sliders and switches that govern every aspect of the Mac and its behavior. You'll be encountering System Preferences a *lot* as you read this book.

Return to this start screen   *Category view, alphabetical view (choose from View menu)*

System Preferences

- **Apple is obsessed with data privacy.** It's a wonderful thing that the company takes such great care with your information. And it's a far cry from Facebook and Google, whose business model is based on selling data about you to advertisers.

Eventually, though, you may get a little sick of macOS's nagging interruptions, seeking your approval for security-related actions. Are you sure the app you've just opened is allowed to use your camera or microphone? Is it OK for a weather app to know your location? Was it really you who just logged into your iCloud.com account on a different computer?

In the end, though, you may well come to appreciate Apple's zeal. Most people feel better not being tracked.

In macOS Big Sur—especially in the Safari browser, the app that's most exposed to the worst people on the internet—you'll find more privacy and security features than ever.

What's New in Big Sur

Apple calls macOS Big Sur its biggest design overhaul since Mac OS X debuted in 2000. And there's no question about it: Big Sur looks amazing.

It's all about color, corners, shadows, spacing, icons. The whole thing feels airier, lighter, cleaner. For example, menu commands are farther apart, and toolbars use more icons and fewer words. Apple even redid all the little *sounds* of the Mac—starting up, dropping something in the Trash, emptying the Trash, taking a screenshot, copying a file—to make things sound cleaner and more modern.

MACS ON APPLE SILICON

Among the Mac technorati, the June 2020 announcement was a bombshell: that after 14 years of using Intel's chips, Apple would begin making its own processors for Macs, starting with some late-2020 models.

It's not so surprising; Apple has, after all, been making its own processors for iPhones, iPads, and Apple Watches from the beginning. Why not the Mac, too?

The payoffs for Apple: lower costs, and freedom from Intel's occasional delays and security problems. The payoffs for you: faster Macs, lower power consumption, much longer battery life on laptops, and—maybe, just maybe—lower prices.

Now, in theory, none of your existing Mac apps should work. Existing Mac software speaks Intel language, not Apple-silicon language.

But Apple, Microsoft, Adobe, and other software companies are adapting their apps so they'll run *natively* (without translation) on the new chips. For all other apps, Apple has created an invisible software translation layer, called Rosetta 2, that allows all your existing software to run normally.

Switching to a different chip is a huge, daunting, very technical task—for Apple. For you, it's pretty much all great news, unless you work for Intel. ★

All of Apple's app icons now have a consistent size and shape (a rounded-edged square) and use the same color palette. The menu bar, open menus, Dock, and desktop-window sidebars are translucent, so your desktop wallpaper can shine through. (You can turn that off, though, if you like.) "Sheets" (miniature dialog boxes) and messages now appear in the middle of the app window, in a clean new design.

The new sheets

Sidebars are a nearly universal element in Apple's apps now—tall, vertical lists of options at the left side of every window. And in Big Sur, they go all the way to the "ceiling" of the window, to show more stuff in less space.

But Big Sur comes with new features, too! Here's the executive summary:

- **Control Center.** Instead of burrowing into a whole different app (System Preferences) to make common settings changes, you can now click the icon on the menu bar to pop up a handy, customizable panel of tiles for Wi-Fi, Bluetooth, Sound, Do Not Disturb, music playback, and so on. It's exactly the same idea as the Control Center on the iPhone, except ... now it's on the Mac.

- **Safari,** the Mac's web browser, is now faster; Apple says frequently visited websites appear twice as fast as they do in Google's Chrome browser. Safari uses less battery power, too, which might be of interest to the 80% of Mac fans who use laptops.

  When no website is open, there's a new start page you can tailor with starting points like favorite websites, your reading list, websites you've been looking at on your other Apple gadgets, Siri suggestions, and so on. And you can give this page its own wallpaper.

More good stuff: When a foreign-language site pops up, you can translate it in place with one click. When you point to a tab without clicking, you get a pop-up preview of the corresponding page. You can now control extensions (plug-ins) on a site-by-site basis.

Above all, though, Apple has put work into security and privacy features. You can now view a privacy report for every website you visit, letting you know just how many scammy cross-site trackers each page has tried to clip onto you (which Safari has blocked). Safari also alerts you if one of your passwords was involved in a corporate data breach, so you can change it before anything bad happens.

- **Messages,** the Mac's texting app, has been remade in the image of its iOS sibling. Full-screen animations, like confetti or fireworks, are now easy to send when you have a message that's full of feeling. You can now create Memoji (giant emoji that look like Fisher-Price Weeble versions of you) right on the Mac. You can pin the icons of up to nine correspondents to

THE BASIC BASICS

Maybe you're new to the Mac. Maybe, in fact, you're new to *computers*. Just to make sure you're covered, here are the terms this book assumes you know:

- **Apps** are, in theory, the reason you bought a Mac. That's the modern word for *software programs*. Safari is an app. Mail is an app. Even Microsoft Word is an app—just a really big one.

- **Clicking.** While the arrow cursor is pointing to something on the screen, press and release the clicker (the mouse button or, on a laptop, the trackpad). You'll do that every time you see a button to choose (like **Save** or **OK**), or when you want to select one icon for copying.

- **Double-clicking.** Without moving the cursor, click twice fast. That's how you *open* something on the desktop, like an app or a document.

- **Dragging.** Move the cursor while holding down the mouse button or keeping your finger on the trackpad. That's how you move an icon or choose from a menu.

- **Shift-click.** While pressing the Shift key on the keyboard, click. You may also be instructed to ⌘-click or Control-click something. Same technique, different keys.

- **Opening a menu.** The menus are the words (usually verbs) at the top of your screen, like , **File**, **Edit**, and so on. Click to reveal a choice of commands.

- **Icons** are the little pictures that represent files, folders, apps, Trash, disks, and everything else on the Mac. When you click one, it darkens to show that you've selected it, in readiness for copying or deleting. ✦

the top of the window, so you don't have to search through the conversations list every time you want to text those people.

You can now choose from a searchable catalog of animated reaction GIFs—those short clips from movies or TV shows, in which a character has exactly the right expression for what you want to say. For the first time, you can reply to individual texts that have already gone by—not just the latest one; these *inline replies* appear indented and attached to the earlier text.

> **NOTE:** Most of these enhancements are for use exclusively between Macs, iPhones, and iPads—that is, they are iMessages features that don't work when you're texting Android phones.

Group chats are better, too. Now you can add a photo for your group, and you can direct texts to individual members of the group.

Best of all, Apple has finally fixed search. You can search for past text messages (and attachments) without crashing Messages.

- **Revamped notifications.** Notifications, the little warnings and alerts that pop onto the upper-right corner of your screen, now work more like they do on the iPhone. Multiple notifications from a single app politely cluster into stacks to save screen space. You can respond to more kinds of notifications directly from their bubbles (not just for Mail and Messages notifications, but also for Podcasts and Calendar).

And now there are mini-windows called *widgets* below the notifications, offering at-a-glance updates on weather, stocks, news, photos, and so on. You can control which ones appear, specify their size and position,

---

THIS BOOK'S ARROW SHORTHAND

As a side effect of macOS's growing complexity, it now takes more steps to find things.

Over and over again, the full instructions to find a certain setting might be, "Click the menu; from that menu, choose **System Preferences**. Now click the **Accessibility** icon. On the Accessibility screen, click **Display**. On the **Display** pane, click **Cursor**. Now you can drag the **Cursor size** slider to make your arrow cursor bigger."

That's a lot of verbiage. So in this book, you'll see a shorthand that looks like this: "Adjust the cursor size in **System Preferences→Accessibility→Display→Cursor→Cursor size**."

Here's hoping you can interpret that notation. Without it, *this* book would be 850 pages long. ✦

---

and download new ones from other software companies. (The Dashboard that used to display these widgets is finally gone for good.)

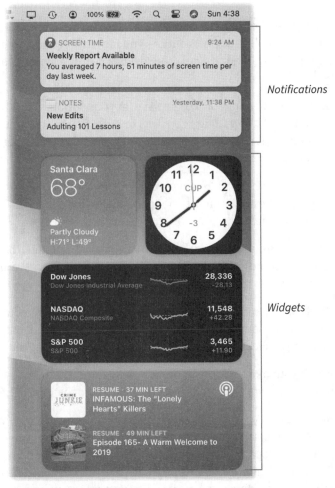

Notifications and widgets

- **Maps** comes with built-in *guides* (folders full of selected sites and restaurants) for a few beloved cities, courtesy of publishers like AllTrails and Lonely Planet. You can also create your own guide folders, full of attractions you encounter as you research a place you're going to visit.

Look Around is now on the Mac for some cities. It's like Google Street View, in which you can "look around you" to see a seamless, street-level photographic representation of a place.

If you travel by bike or electric car, you'll find that Maps' routing includes useful information like elevation, hills, and charging stations. There are even maps of *indoor* places like airports and malls.

- **AirPods switching.** If you own AirPods (Apple's wireless, detached white earbuds), you'll enjoy this one: As you move from one Apple machine to another (Mac, iPhone, iPad), the AirPods switch automatically, too.

- **Battery.** New graphs show what your laptop battery has gone through in the past 24 hours or 10 days. And, at your option, you can turn on Optimized Battery Charging, a way to make your battery last more years by charging it only to, for example, 80% each day. (Full charging is bad for lithium batteries over time.)

- **FaceTime.** When you're on a group call and somebody starts using sign language, the app is smart enough to make that person's video box big enough to see.

- **Security cameras.** If you have a home security camera that uses Apple's HomeKit standard, it can now learn to identify faces, so you won't be alarmed when a family member comes to the porch.

- **Photos.** You know all those editing tools for photos, like Rotate, Crop, color corrections, and filters? Now you can apply all that to videos, too. Finally, you can fix that iPhone video that got recorded sideways.

  And if you take the time to name your photos, the descriptions you type (now called *captions*) sync, so they'll appear on your iPhone and iPad, too.

- **Reminders.** Apple's to-do app has been getting awfully robust in the past few versions, but now it's even robustier. For example, you can split up tasks by delegating them wirelessly to other people, who receive notifications of their new burdens. There are new keyboard shortcuts, new automated reminder suggestions, and smarter smart lists.

- **Spotlight,** the Mac search feature, presents a very different results list now. The preview pane doesn't appear until you click one of the results— and when it does appear, you can now rotate, crop, and annotate graphics and PDFs right in that Preview window. Meanwhile, the results list follows a very new logic, described on page 121.

- **Voice Memos.** This sound-recording app now lets you organize your recordings into folders, mark important ones as favorites, and remove background noise or room echo.

- **Battery time remaining is back!** Yes, Apple, we know—it's an estimate, and it could rise or fall depending on how hard we're driving our laptop.

But how great to get a rough idea of the time left before battery depletion, just by clicking the ▬▸!

- **The startup chime is back!** When you turn on your Mac, it once again makes the majestic F-sharp major chord that graced our ears from 1998 to 2016, when Apple mysteriously axed it.

  Note to librarians and churchgoers: Not to worry. If you've muted your Mac, the startup chime doesn't play. You can also turn it off in **System Preferences→Sound→Sound Effects→Play sound on startup**.

With this quick summary, the following pages, and an optimistic attitude, you should have no problem diving into Big Sur—and unlocking your Mac's full potential.

PART ONE

Meet the Machine

CHAPTER ONE

The Mechanics of the Mac

I n 1997, Apple co-founder Steve Jobs returned to the company that had fired him in 1985. One of his first actions was to clean house—to simplify the assortment of 47,000 different Mac models the company was then making. He decided that from that moment on, Apple would manufacture only four Macs: two each for the desktop and the laptop.

The 1997 Mac lineup

Today, of course, Macs look a lot better, run a lot faster, and incorporate a lot less candy-colored plastic. Each model now comes in several sizes and speeds, and the oddball, screenless Mac mini doesn't really fit into the lineup. But the spirit of the original 1997 product grid lives on.

Fortunately for you (and for computer-book authors), they all contain exactly the same components. There's always a screen, a keyboard, a speaker, a power cord, and so on. And they all run the same operating system: a huge glob of software and apps (programs) called macOS. (It's

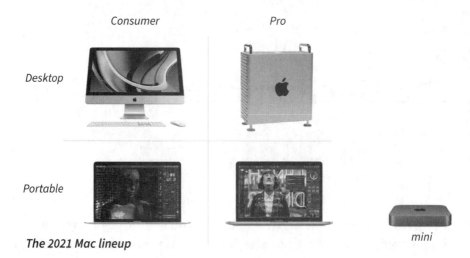

	Consumer	Pro
Desktop		
Portable		mini

The 2021 Mac lineup

pronounced "mac oh ess." Don't embarrass yourself at a party by saying "mac-oss.")

The subject of this book is the 2020 version of macOS: version 11.0, which Apple has nicknamed Big Sur. But before you start exploring the nooks and crannies of the Mac's *software*, here's a guided tour of its *hardware*.

On, Off, and Asleep

When your Mac comes from the factory, it's turned off; Apple doesn't want it to arrive with a dead battery. You turn it on like this:

- **Laptop.** On a MacBook Whatever, press the top-right key. On recent models, you may not even recognize it as a key, because it's a dark, flat

It's the fingerprint reader, too.

The MacBook power key

square without a label, and it doubles as the fingerprint reader. But it's there, top right.

- **Desktop.** If you are the proud owner of an iMac, Mac Pro, or Mac mini, press the ⏻ button on the back or top of the computer.

> **NOTE:** If you are one of the 11 people who bought the rack version of the Mac Pro—the one that's horizontal, intended to be bolted into a rack in some humming underground data center, and whose base model starts at $6,500—you're probably a network administrator. Therefore, you probably know already that the power button is an unlabeled, featureless capsule-shaped button on the back panel. Apple didn't want to insult you by painting the ⏻ label on it.

That ecstatic moment of unboxing a new Mac may be one of the *only* times you ever use the power button. For the rest of its life, when you're not using the Mac, you aren't supposed to turn it off. You're supposed to *put it to sleep*.

If you have a laptop, just close the lid to induce sleep. On a desktop model, choose **⌘→Sleep**, or just wait awhile. After a few minutes of activity, the screen goes black, and the computer dozes off.

> **TIP:** You can control how quickly the Mac goes to sleep in **System Preferences→ Battery** (on laptops) or **System Preferences→Energy Saver** (on desktop Macs).

When the Mac is asleep, its screen goes black, and it uses very little power. But all your work, in all the open apps, is actually still in the computer's memory.

As soon as you wake the machine up again—by pressing any key or opening the laptop lid—the screen lights up with everything exactly the way it was.

The lesson here: When you finish a work session, let the Mac sleep instead of shutting it down. It costs you almost no electricity, and next time you want to work you'll save time.

> **NOTE:** Even when the Mac is asleep, it still performs internet tasks like downloading email and synchronizing files on your iCloud Drive or Dropbox—as long as the Mac is plugged in.
>
> This feature is called Power Nap. If you're crazy, you can turn it off in **System Preferences→Battery→Power Adapter** (laptops) or **System Preferences→Energy Saver** (desktop Macs).

The Mouse/Trackpad

In one important area, your Mac is not like an iPad or an iPhone: It doesn't have a touchscreen. Apple's thinking goes like this: Sitting in a chair with your arm stretched out toward a vertical screen, for hours a day, is a recipe for an injury whose name—"Gorilla Arm"—is much more fun than its feeling.

So don't try to reach out and touch something. Instead, you'll use either the mouse or the trackpad, just as computer buffs have since the Reagan administration.

- **The mouse**—Apple's version—is a white capsule about the size of a bar of soap. It's available in both wireless and corded versions. You roll it across the desk to move the cursor on the screen.

> **TIP:** You can control how far the cursor moves on the screen, relative to how fast you move the mouse. Visit **System Preferences→Mouse**, and adjust the **Tracking Speed** slider.

- **The trackpad,** built into Mac laptops (and available as an add-on for desktops), lets you move the cursor by sliding your finger across the pad. But it also permits all kinds of other stunts. For example, you can scroll through a document or a web page by dragging two fingers on the trackpad. And in some apps, you can actually make *drawings* using the trackpad as your canvas.

Shortcut Menus

Lurking behind almost everything on the screen—every file, folder, typed word, picture, web page, or whatever—is a hidden menu of things you can do to that thing. It's a short menu of actions like **Rename, Copy,** or **Move to Trash.**

At the dawn of computing, these shortcut menus were known only to the geeky intelligentsia. Apple hid them because they contained technical options intended only for nerds.

Today, though, shortcut menus can save you a lot of time and fumbling, and sometimes they contain important options that aren't available anywhere else. Next time you're trying to accomplish something in some app, flailing and lost, say to yourself, "Oh, right! Maybe the command I want is in the shortcut menu!"

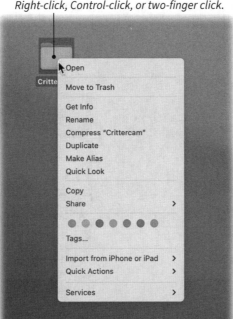

Right-click, Control-click, or two-finger click.

The secret shortcut menu

Because this secret menu is so important, Apple has provided an absurd number of different ways to open it. For example:

- **Control-click.** That is, while pressing the Control key (on the bottom row of your keyboard), click the mouse or trackpad on your target. The shortcut menu appears at the tip of your cursor.

- **Right-click.** There's a one-handed way to open the shortcut menu, too: Click something on the screen by pressing the *right* mouse button.

  Your Apple mouse actually has *two* side-by-side buttons. It sure doesn't look that way—but under that smooth white plastic, there are in fact two different places to press. But the right button doesn't work until you ask for it, as described in "Unlocking the right mouse button" (next page).

  (Why does Apple call it "secondary click" and not "right click"? Because many left-handers reverse the functions of the left and right buttons.)

> **TIP:** You may be the proud owner of what Apple calls a Magic Mouse (or Magic Mouse 2). It's a sleek, sculpted, flattened mouse whose top surface acts like a trackpad. You can swipe your finger across it to scroll. You can tap, double-tap, tap with two fingers, and otherwise use it much as you would a trackpad. You'll find more about the Magic Mouse throughout this book.

In System Preferences→Mouse, turn on this checkbox.

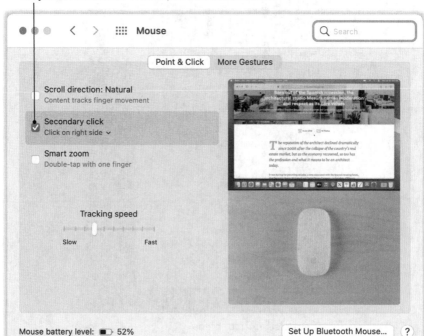

Unlocking the right mouse button

From now on, your Mac registers a regular click (left-click) or a right-click (to open a shortcut menu), depending on which side of the button you press.

> **TIP:** If this business of invisible right and left buttons freaks you out, you may prefer a mouse with two clearly separated buttons, like the ones on Windows PCs. Fortunately, any old $10 USB mouse works when plugged into your Mac, even if it was originally designed for a PC.

- **Two-finger click the trackpad.** On your laptop trackpad, the shortcut menu pops right up wherever you click with *two* fingers.

> **TIP:** In **System Preferences→Trackpad→Point & Click**, you can set things up so clicking in a corner of the trackpad—with only one finger—opens the shortcut menu.

In this book, you'll find the instruction to "right-click or two-finger click" whenever you're supposed to open a shortcut menu.

The Keyboard

As you may have figured out from its price tag, the Mac is not a typewriter. Its keyboard has a lot more than 26 letter and 10 number keys—and plenty of the extra keys aren't what you'd call self-explanatory.

Mac keyboards have evolved over the years, but if yours was made since 2003 or so, it has this set of bonus keys. Most of them are on the top row:

- **☀, ☀ (F1, F2).** These keys adjust the brightness of your screen.

> **NOTE:** The next few paragraphs describe the keys at the very top of your keyboard, labeled with both logos (☀, ☀) *and* names like F1 and F2. On MacBook Pro laptops, you'll find a narrow touch strip instead of these physical keys. It's called the Touch Bar, and it's described on page 73.

- **▦ (F3)** opens a view called *Mission Control*: All windows in all open programs shrink down to miniatures, so you can click to open the one you're looking for in the stack of overlapping windows.

Every window in every app gets its own thumbnail, so you can see what's running.

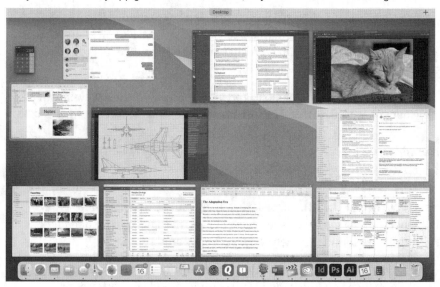

Mission Control

- **⊞ (F4)** opens the *Launchpad*: a screen full of all your apps, arrayed for easy clicking. (Older keyboards might have a ⊙ key instead.)

Launchpad

- **⌁, ⌁ (F5, F6).** On most Mac laptops, the keys automatically *light up* when you're typing in a dark place. But these keys let you adjust their brightness manually. (On desktop Macs, the F5 and F6 keys don't have any preassigned function at all.)

- **◄◄, ►‖, ►► (F7, F8, F9).** In music and video programs, these keys mean just what you would expect: Rewind, Play/Pause, and Fast-Forward.

> **TIP:** *Tap* the ◄◄ or ►► key to skip to the previous or next track. *Hold it down* to rewind or fast-forward.

- **◄, ◄), ◄)) (F10, F11, F12).** These keys adjust your speaker volume. Tap ◄) repeatedly to make the sound level lower, ◄)) to make it louder. The ◄ key mutes the sound completely; tap again to unmute.

  With each tap, a big gray ◄) or ◄)) floats on your screen for a couple of seconds, just so you know the Mac has understood your intentions.

- **⏏.** This is the Eject key, available only on desktop keyboards. You're never supposed to yank out a flash drive, disk, or a memory card without giving the Mac a heads-up. The risk is that you might interrupt some file-saving process, losing data.

That's the purpose of the ⏏ key: When there's a flash drive, memory card, CD/DVD, or external hard drive connected to your Mac, you can click its icon and then press this key. (If you don't have the ⏏ key on your keyboard, you can use the **File→Eject** menu command instead.) The disk or card icon disappears from the screen. *Now* it's safe for you to pull it out or disconnect it physically.

- **Delete** is your backspace key.

- **Return (or Enter).** When you're typing, you press Return to begin a new paragraph. The rest of the time, you can use it to mean **OK** or **Save** or **Print** or **Done** when a dialog box is on the screen, as described on page 4.

- **Esc.** There's a reason the Escape key occupies such a prominent place at the upper-left corner of your keyboard: because you'll use it all the time once you get to know it.

  Escape means "cancel." You can use it to close a dialog box, back out of a menu, end a slideshow, exit full-screen mode, dismiss your screen saver, and so on.

Modifier Keys

Some keys don't do anything when you press them by themselves. They exist only to change the effect of *other* keys.

You may be familiar with the Shift key, for example; its purpose is to capitalize whatever letter key you strike. Shift is therefore an example of a *modifier key*—but it's not the only one you have:

- **⌘.** The ⌘ symbol, believe it or not, means "special feature" in Swedish hiking guidebooks. But when you're not hiking in Sweden, this incredibly

---

A CRASH COURSE IN KEYBOARD COMBOS

In this book, and wherever fine Mac tutorials are found, you'll encounter instructions like "Press ⌘-B to make the text bold."

What you're being asked to do requires two fingers—sometimes two hands. While you're holding down the ⌘ key with one finger, type the letter key (B in this example) with the other. Now

release both keys. This technique quickly becomes second nature.

And why would you bother with learning a keyboard shortcut for a command? Because the alternative is to take one hand off the keyboard, grab the mouse, and click some button or menu on the screen. That's a lot of effort when you're in the middle of a burst of inspiration. ★

---

important key lets you use menu commands without removing your hands from the keyboard to grab the mouse.

For example, when you're ready to quit the app you're using, you can press ⌘-Q instead of using the menus.

- **Control (⌃).** You probably won't use this key much, but it's one way to "right-click" to open shortcut menus, as described on page 20.

- **Option (⌥).** The Option key (sometimes labeled "Alt") is a modifier key without a consistent function. Among other things, it helps you type special symbols like ¢ (Option-4) and © (Option-G).

- **fn.** Apple presumably intends for you to pronounce this as "function," because it changes the *function* of certain other keys. That's especially useful on laptops, which don't have enough room for every key you might hope for.

For example, the Delete key ordinarily backspaces, deleting whatever is to the left of the blinking cursor. But if you're holding down the fn key, the Delete key becomes a *forward*-delete key, deleting whatever is to the *right* of the cursor.

KEYS ON THE EXTENDED KEYBOARD

Most modern Mac keyboards no longer include a number keypad off to the right. Apple decided that a block of duplicate number keys that hardly anybody uses wasn't worth the space, much to the disappointment of the world's accountants.

But if you are among the few whose keyboard includes a number pad, then you get a bonus block of keys *between* the main keyboard and the numbers:

- **Home** returns you to the top of a document, window, list, or web page; the **End** key to the bottom. (On keyboards without a number keypad, you can trigger the Home and End functions by holding down **fn** as you tap the ◀ and ▶ arrow keys.)
- **Page Up** and **Page Down** scroll your window or web page up or down by

one screenful. Super useful, really. (Here again, you can simulate these functions even if you don't have those special keys: Press **fn** plus the ▲ and ▼ arrow keys.)

- **Clear** removes from the screen whatever you've highlighted, but without putting a copy on the invisible Clipboard, as the Cut command would do.

- ⌦ is the forward-delete key. It deletes whatever is just to the *right* of the blinking insertion point, like a reverse Backspace.

- Finally, the **Help** key opens the help screens for macOS, Microsoft programs, and certain other apps. But you probably didn't need any help to guess that one. ★

The Screen

You might not think there's much worth saying about the screen. It's a screen. You look at it. In fact, though, you can make your screen easier to see, easier on your eyes, and friendlier for letting you fall asleep at night. For example:

- **Adjust the brightness.** You already know you can make the screen brighter or dimmer using the ☀ and ☀ keys. But out of the box, the Mac attempts to adjust the brightness automatically, based on the brightness of the room. (Yes, it actually has a light detector.) That's nifty, but it can also be annoying if you prefer to have manual control.

  The on/off switch for this feature is in **System Preferences→Displays**. It's called, of course, **Automatically adjust brightness**.

- **Make everything bigger.** Getting older means accumulating wisdom and life satisfaction, but it can also mean your vision starts to go. Fortunately, the Mac can enlarge everything on your screen with just a couple of clicks.

  We're not talking about just making the *text* bigger, for reading email messages or web pages—see pages 174 and 197 for that. We're talking

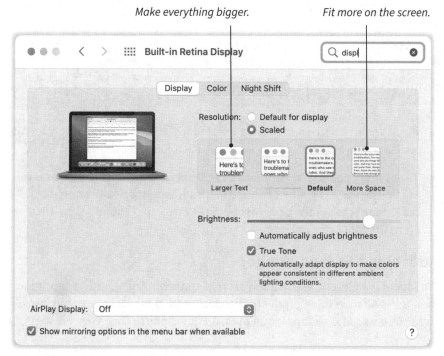

Changing your screen resolution

about enlarging *everything* on the screen, zooming into it. Of course the trade-off is that not as much will fit on the screen at once.

To make it so, choose **System Preferences→Displays→Display**. The setting called **Default for display** provides the sharpest but smallest image. If you choose **Scaled** instead, you get to see a list of other screen resolutions, represented either by pixel dimensions (like **1440 × 900**) or by little thumbnails, showing the effect each choice will have on the image.

- **Fix the color.** Photographers sometimes refer to the *color temperature* of a scene—the tint of the light. To your brain, a white sheet of paper always looks white, whether it's under cold fluorescent light or warm sunset light. But if you could see photographs of the two scenes side by side, you'd see a huge difference.

  The Mac can adjust the tint of the colors on its screen automatically, mimicking the way your brain works, so the colors seem consistent from one lighting condition to another. This feature is called True Tone, and the on/off switch is in **System Preferences→Displays→Display**.

- **Accommodate your melatonin.** In the tech industry, there's a common belief that the bluish light emitted by computer screens disrupts your brain's production of melatonin, the "I'm getting sleepy" chemical, making it harder to fall asleep.

> **NOTE:** There's not *really* much research that backs up that claim. More likely, it's the stimulation of whatever you're watching or reading that makes it hard to fall asleep. In fact, the best way to get sleepy is to just stop *using* your screens a couple of hours before bedtime. But let's play along.

If you wish, the Mac can begin adjusting its colors to become warmer and less blue as your bedtime approaches. It's a feature called Night Shift.

To set it up, open **System Preferences→Displays→Night Shift**. Here you can set up the warming effect feature to kick in automatically when it gets dark (**Sunset to Sunrise**), on a schedule you set up (**Custom**), or manually (**Turn On Until Later Today**).

Attaching a Second Monitor

For lots of people, one Mac screen just isn't enough. Connecting a second monitor is an incredible productivity booster. On one screen, you can keep, say, your email and chat program open all day. On the other, you can see whatever real work you're doing—maybe a spreadsheet or a video you're editing.

Or maybe you want the two screens to show the *same* image. That might be handy when you're trying to display a slideshow in front of an audience, on a projector or TV you've connected to your Mac.

To plug in a projector or second screen, investigate the kind of cable it has. The most common kind is called an HDMI cable, which carries both the video and the audio from the Mac to the monitor. Unfortunately, most Macs don't have HDMI jacks, which means you'll need to order the right kind of adapter from someplace like Apple (expensive) or Amazon (cheap). Most modern Macs have either what Apple calls Thunderbolt 2 or Thunderbolt 3 (USB-C) jacks. In other words, you're shopping for, for example, a "Thunderbolt 3 (USB-C) to HDMI adapter."

Thunderbolt 2 on an iMac

Thunderbolt 3 on a MacBook Pro

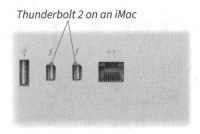 

Thunderbolt 2, Thunderbolt 3

The Mac can also throw its audio and video signals *wirelessly*—if you own an Apple TV (a little black box that connects to your TV and feeds it all the internet video services like Netflix, Hulu, and so on).

In any case, once everything is connected, open **System Preferences→ Displays→Arrangement**. Here you see a miniature image of each monitor. You can drag these images around within the window, thereby teaching the Mac where the second monitor *is* relative to the first one.

Usually, people put the second monitor off to the right. That way, when you move the arrow cursor to the right edge of the main Mac screen, it moves smoothly onto the left edge of the second screen.

Here, too, is the **Mirror Displays** checkbox. If you turn it on, the Mac shows the same thing on both screens, rather than treating the second monitor as an extension of the main one.

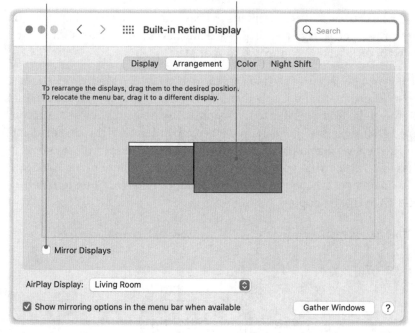

Turn on to make the two screens look identical.

Drag this miniature so it matches your second screen's real-world position.

Arranging a second monitor

The iPad as a Second Screen

Believe it or not, you may already own a second monitor for your Mac. It's called the iPad.

That's right: If you have a recent model iPad (2018 or later), you can set it up as an external monitor for your Mac, thanks to a feature Apple calls Sidecar.

> **NOTE:** The Mac has to be fairly recent, too. It must be a 2016 MacBook, MacBook Pro, or iMac or later model (the iMac 5K works, too); a 2018 or later MacBook Air or Mac mini; or a 2019 or later Mac Pro.

You can connect the iPad to the Mac with a cable (a Lightning cable or USB-C cable, depending on your model). That way, the iPad stays charged up. But you can also connect the Mac to the iPad wirelessly. Both machines have to be signed into the same iCloud account (page 150).

Suddenly, a new icon appears on your Mac menu bar, in blue, and in your Control Center: ▣. Click it and then choose the iPad's name.

Drag app windows right off the edge of the Mac screen, through space, onto the iPad.

Sidecar: iPad as second screen

The iPad's screen lights up as an extension of the Mac's screen. You can move your cursor from one to the other, as though they were connected.

> **TIP:** If you prefer, the iPad can also show a duplicate of what's on the Mac screen. To set that up, open the ■ menu. Choose **Mirror Built-In Retina Display** (or whatever your Mac's screen is called). To switch back, use the same menu to choose **Use As Separate Display**.

In Sidecar, the iPad can be more than just a second screen. It can also act as a limited touchscreen stand-in for your Mac:

- **Draw, point, and click with the Apple Pencil.** Apple sells a $130 plastic stylus called the Apple Pencil. If you have one, you can use the iPad as a graphics tablet—you can draw on the iPad screen directly into Mac graphics apps like Photoshop, Illustrator, Final Cut Pro, and Maya.

- **Scroll the iPad screen.** Drag up or down with two fingers.

- **Use the Sidecar sidebar.** The sidebar is a thin column of buttons on the iPad screen that offer some control over the Mac. If you're sitting with the iPad in your lap, the sidebar lets you perform some Mac tasks. It contains buttons for Undo, Show Keyboard, modifier keys like ⌘ (Command), ⌥ (Option), ⌃ (Control), and ⇧ (Shift), and so on, which are frequently important in graphics programs.

- **Use iPad apps.** Even in Sidecar mode, you can still swipe up from the bottom of the screen to reveal the iPad's Dock as usual. It gives you access to any of your iPad apps.

When you're finished using the iPad as a monitor, choose Disconnect from the ▣ menu (on the Mac), or tap ⬚ (on the iPad's sidebar).

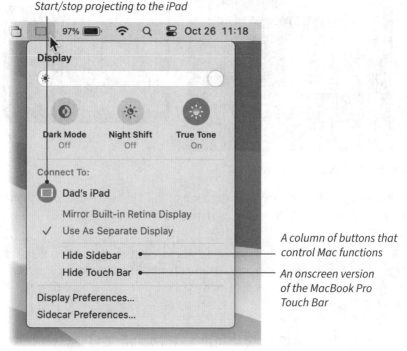

Start/stop projecting to the iPad

A column of buttons that control Mac functions

An onscreen version of the MacBook Pro Touch Bar

Sidecar setup

Speakers and Microphones

You already know about the ◂) and ◂)) keys, which adjust your Mac's speaker volume, and the even more essential ◂ key, which silences it instantaneously. But what are you supposed to do if a rabid rottweiler bursts into the room, snarling and snapping, irrationally attacks your keyboard, and reduces it to a pile of foamy plastic splinters? How will you adjust the volume now, huh?

Fortunately, you can also use the volume slider in the Control Center, which appears when you click the ◂)) on the menu bar. (See page 40 for more on the Control Center—and how it lets you put the ◂)) volume control back onto the menu bar, where it always sat before macOS Big Sur came along.)

The Mac can also accommodate *more than one* gadget for bringing sound in and out of your Mac. For example, your Mac has speakers, but you may also want to plug in a pair of headphones. How do you tell it whether the sound should play through the speakers or the headphones? Or maybe it has a built-in microphone, but you've hooked up a fancier external USB microphone that you bought for making podcasts. Which one should the Mac listen to?

You can make those decisions in the Control Center, too. Click anywhere on the Sound tile *except* on the slider to open the panel shown in the figure "Volume and input/output controls in the Control Center." It lists, under **Output**, all the different ways your Mac can play sound—speakers, headphones, wireless amplifiers—so you can click the one you want.

If you Option-click the Sound tile instead, you also get an **Input** section, which lists all the different microphones you've got connected. Click the one you want on duty.

> **NOTE:** Just to make sure you don't miss the point, Apple includes yet another set of the same controls in **System Preferences→Sound**.

AUTO-SWITCHING AIRPODS—AT LAST

If you're not a devotee of AirPods (Apple's completely wireless, completely detached earbuds), then Big Sur's crowning achievement may not mean much to you. But here it is: When you've been listening to audio from your iPhone or iPad, and then sit down at the Mac, the AirPods switch automatically. They start playing the Mac's audio for you, without your having to fiddle with any menus or change any settings.

Well, that's what Apple says; it's not actually *quite* that simple. The first time you enter the range of another Apple gadget that's on your same iCloud account, you get a notification that says, "AirPods Nearby." You have to point to it and click **Connect** to complete the switch.

At this point, your sound menulet (usually ◀)) changes to look like a pair of iPods. You can use this menu to confirm the connection, check the Pods' battery

level, or, on the Pro models, switch their noise-cancellation setting.

To disconnect, choose a different speaker from that menu—or just put the AirPods back into the case. ★

Option-click Sound...   *...to see your list of audio inputs.*

Volume and input/output controls in the Control Center

What to Plug In Where

The last stop on this tour of your Mac is the set of jacks, or ports, on the back or the side. The assortment varies depending on the model of Mac you have, but it always includes some combination of these:

- **USB.** This thin, rectangular jack may be the most commonly used connector in all of computerdom. Into it, you can plug printers, scanners, cameras, flash drives, DVD drives, hard drives, Ethernet network adapters, and so on.

  Because the USB jack supplies power, this is also where you plug in the charging cables for iPhones, iPads, fitness bands, and so on.

- **USB-C.** On some recent models, Apple has replaced (or complemented) the traditional USB jack with a more modern one called USB-C. In theory, USB-C is fantastic.

  This tiny connector can carry power, video, audio, and data simultaneously. It can replace your Mac's power cord, USB jacks, video output jack, and headphone jack. (On MacBook Pros, for example, you can plug the power cord into any of the four USB-C jacks.)

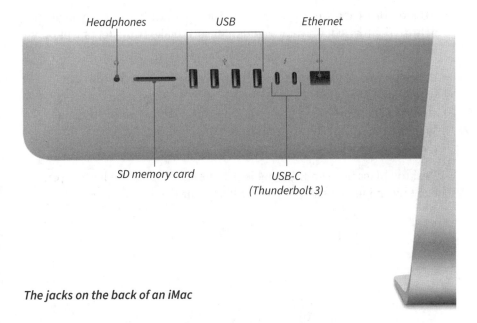

Headphones USB Ethernet

SD memory card USB-C
(Thunderbolt 3)

The jacks on the back of an iMac

A USB-C cable is identical top and bottom, so you can't insert it upside-down, as you can with traditional USB. USB-C can charge external gadgets faster, and transfer data to flash drives or hard drives faster, too.

And the brand doesn't matter. Your friend's Samsung USB-C cable can charge your Apple MacBook and his Surface tablet. You're witnessing the dawn of the universal charging cord!

But change is annoying, and the world didn't switch over to USB-C as fast as Apple had hoped. Of course, you *can* buy flash drives, hard drives, cameras, and charging cords that connect with USB-C. (The hard drives are the most attractive, because Apple's version of USB-C, which it calls Thunderbolt 3, transfers data much, *much* faster than traditional USB.)

But it still seems as though 95% of the gadgets you might want to plug into a Mac stubbornly come only in traditional USB format. To make them work in a USB-C jack, you have to attach a $5 adapter. You'll get to know those adapters really well.

A USB-C adapter

- **Headphone jack.** Every Mac still comes with a traditional headphone jack. It can accommodate the standard miniplug on any headphones, or even the white earbuds from an iPhone or iPad.

- **SD memory-card slot.** If you're a photographer with a laptop, you must ache to see this slot on your friends' iMacs. It makes transferring photos and videos from a camera fast and effortless.

- **Ethernet jack.** Most computers these days connect to the internet wirelessly, using a Wi-Fi connection (page 329). But you can still plug a physical network wire—known as an Ethernet cable—into your Mac, thanks to this jack. The advantage: better speed than Wi-Fi, and no possible way for bad guys to eavesdrop on what you're doing.

  Desktop Macs have an Ethernet jack built in. If you have a laptop, you can buy an inexpensive adapter (USB or USB-C) that provides one.

CHAPTER TWO

Sixteen Settings to Change Right Now

Apple's goal is to create a single computer, running a single operating system, that serves every conceivable customer. Its dream would be to make everybody equally happy: 9-year-olds in Paraguay, mathematicians at SpaceX, Wall Street bankers who do yoga, octogenarian rock musicians, one-armed chemists, Irish computer science professors, Indonesian nuns ... and you.

Unfortunately, it can't be done. Everyone is different. The design and settings that seem perfect to someone else might seem unusable to you.

All Apple can do, therefore, is make sure you can *change* every one of the thousand controls, buttons, sliders, and switches in System Preferences (the Mac's settings app)—and then guess which factory settings will please the most people most of the time.

That's why, upon installing macOS Big Sur or getting a new Mac, you should take a few minutes to change some of the settings to make the Mac *your* Mac. This exercise will introduce you to the locations of some of the Mac's most important control centers.

To make most of these tweaks, you'll begin by opening System Preferences. You can choose its name from the menu at the top-left corner of your screen or click its icon on the Dock (page 85).

Fix the Scrolling Direction

For many years after the invention of the laptop, nobody questioned how scrolling worked on the trackpad. When you drag two fingers *down* the trackpad, you scrolled *downward* in the document—toward the end of it.

But somewhere along the line, somebody at Apple said, "Hey, wait a minute. On the iPhone and the iPad, scrolling with your finger works the opposite way! When you drag downward on the screen, you're scrolling upward, toward the beginning of the document! We've had it backward all along!"

And so, one fine morning in 2011, Mac fans discovered that Apple had swapped the direction of trackpad and Magic Mouse scrolling. Suddenly, dragging *downward* on the Mac trackpad scrolled their windows *upward*.

If you're not used to it—for example, if you're accustomed to the way scrolling works on Windows laptops—it'll make you crazy.

Fortunately, you can bring back the traditional down-is-down scrolling system easily enough. Open **System Preferences→Trackpad**, click **Scroll & Zoom**, and *turn off* **Scroll direction: Natural**.

As you do so, you're allowed to mutter, "Maybe it's natural to *you*, Apple!"

Bring Back the Scroll Bars

If everybody would just limit their emails to one paragraph, edit down their Word documents to a couple of good points, and constrain their web articles to 500 words, we wouldn't need *scrolling*. Everything would fit onto one screen, and everybody would be happy.

In the real world, of course, your screen isn't tall enough to contain everything anybody writes or creates in a single view.

That's why scroll bars were invented. In any window that's too small to show everything at once, you get a vertical strip at the right edge. Inside that strip,

---

EVEN MORE WAYS TO SCROLL

Because scrolling is such an important part of using a computer—it's something you do hundreds of times a day—your Mac comes with a lot of different ways to do it.

When you're looking at a web page or reading an email, *press the space bar* to scroll down exactly one screenful. That's a fantastic, efficient trick you should memorize before the day is out. (It's much less common that you need

to scroll *up* by one screenful—but when the time comes, you can do that by holding down the Shift key as you tap the space bar.)

In any app, you can press the Page Up or Page Down keys to scroll by one screenful. Very few keyboards actually have these keys anymore, but you can also simulate their functions by pressing **fn** plus the ▲ and ▼ arrow keys. ✦

---

like the elevator inside a shaft, is a dark handle. The scroll bar is like a map of the entire window's contents: When the elevator is at the bottom of the shaft, you're at the end of the document; when it's at the top, you're at the beginning.

> **NOTE:** The *size* of the handle changes to illustrate how much of the document you're seeing at the moment. If the handle is half the height of the scroll bar, then you're currently seeing half the window's contents. If the handle is very short, then either your window is very small, or the document you're about to read is going to take you a very long time.

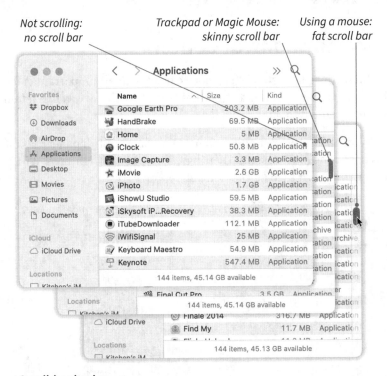

Not scrolling: / *no scroll bar*   *Trackpad or Magic Mouse:* / *skinny scroll bar*   *Using a mouse:* / *fat scroll bar*

Scroll-bar basics

None of that is especially complicated to master—except that, in macOS, Apple makes the scroll bar *disappear* when you're not using it. That's an added obstacle to mastery that nobody asked for.

Fortunately, it's easy to fix. Open **System Preferences→General**. Where it says **Show scroll bars,** you have three options:

- **Automatically based on mouse or trackpad.** If this option is selected, then the scroll bars always appear if you've plugged a *mouse* into the Mac.

If not, then you must be using a trackpad, and Apple figures you'll scroll using the "drag two fingers up or down" technique. Or you might be using Apple's Magic Mouse, whose top surface acts like a trackpad; you can drag *one* finger up or down to scroll. In these cases, the Mac hides the scroll bars except when you're actually in the act of scrolling.

- **When scrolling** means the scroll bars don't appear until you start scrolling. At that point, they show up so you can see where you are in the document, thanks to the little elevator-in-shaft map.

- **Always** means the scroll bars are always visible. For most people learning the Mac, **Always** is the most useful option.

Set Up Your Control Center

Most of the settings in System Preferences, frankly, are unimportant. A few, though, you'll use very often—so often that trundling all the way to System Preferences to adjust them gets old fast.

That's why, in macOS Big Sur, there's a new, compact control panel that offers quick access to the most essential settings, like volume, brightness, Wi-Fi, and Do Not Disturb. It's called the Control Center.

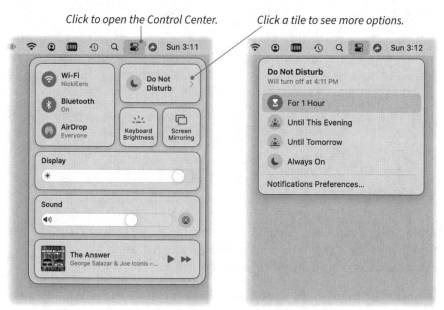

Control Center—and an expanded tile

To open the Control Center, click ⚏ on your menu bar.

The first-level controls are all here—volume, brightness, Wi-Fi, and so on. But if you click a tile (anywhere *except* on an actual slider or button), it expands to reveal even more detailed settings. For example, if you click the **Sound** tile, it expands to reveal your list of microphones and speakers. If you click the **Do Not Disturb** tile, you're offered additional options like **For 1 Hour, Until This Evening,** and **Until Tomorrow.**

And here's the beauty of it: *You* control which controls appear on the Control Center, as described in the next section. Taking the time now to set it up for *your* purposes will pay off in efficiency later.

Design Your Menu Bar

Popping open the Control Center is certainly an improvement over burrowing into System Preferences. But it's still a couple of clicks—one to open Control Center and one to make your tweak. Fortunately, Big Sur offers yet another way to put settings at your fingertips: You can install the really important ones right onto the menu bar. Just drag the tile itself directly up onto the menu bar, as shown on the next page.

Now you've got *direct* access to your sound settings, brightness settings, playback controls, Do Not Disturb settings, and so on.

The dragging thing is fun, but there's actually a more centralized and complete way to design your Control Center and menu bar. That's to open **System Preferences→Dock & Menu Bar** (see page 43).

This panel, newly overhauled in Big Sur, contains a list of every available tile your Control Center offers. Some of them, like **Accessibility Shortcuts** and **Fast User Switching,** don't even come preinstalled in the Control Center!

In any case, you can click each one of these options and then choose, in the main window, either **Show in Menu Bar, Show in Control Center,** or both.

You can drag a tile right out of the Control Center...and up onto the menu bar.

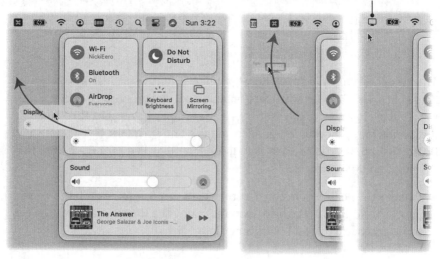

Dragging from the Control Center to the menu bar

Here's a rundown of your options:

- **Wi-Fi.** Shows whatever wireless network you're connected to. Click the tile to see an on/off switch for Wi-Fi, a list of other available networks, and a link to the Network window in System Preferences.

- **Bluetooth.** Shows if Bluetooth is turned on. Click the tile to see an on/off switch for Bluetooth, a list of Bluetooth gadgets you've used with this Mac (like wireless speakers or earbuds), and a link to the Bluetooth window in System Preferences.

- **AirDrop.** Apple's AirDrop feature lets you send files, pictures, and other information between Macs, iPhones, and iPads, wirelessly and without much red tape. You can read more about it in Chapter 14. This tile shows who is allowed to exchange stuff with you using AirDrop. Click the tile to see an on/off switch for AirDrop and a choice of who's allowed to AirDrop with you (**Everyone** or **Contacts Only**).

- **Do Not Disturb.** When you turn this feature on, no notifications or warnings (beeps, chimes, or pop-up bubbles) will interrupt your work, as described on page 48. When the ☾ logo is blue instead of gray, Do Not Disturb is turned on. Click the tile to see a choice of durations for the Do Not Disturb to be in effect, like **For 1 Hour** or **Always On**, and a link to the Notifications window in System Preferences.

- **Keyboard Brightness.** Click this tile to see a slider that adjusts the brightness of the backlighting behind the keys of your keyboard in

Some mini-controls can appear in the Control Center, the menu bar, or both.

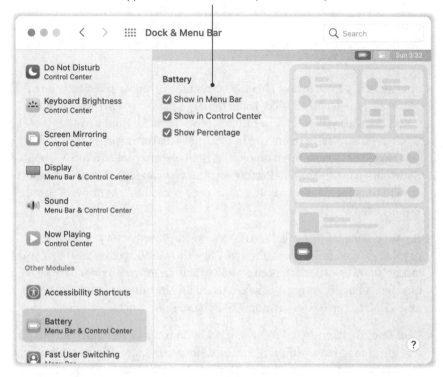

Dock & Menu Bar (and Control Center) preferences

low-light situations. You also get a link to the Keyboard window of System Preferences.

- **Screen Mirroring.** This tile turns blue to let you know you're displaying the Mac's screen image on an external screen or projector, as a warning to be careful what websites you pull up. Click the tile to see a list of available screens or projectors, and a link to the Display window of System Preferences.

- **Display** offers a brightness slider for your screen. Click the tile to see on/off buttons for **Dark Mode** (page 61) and **Night Shift** (page 28), and a link to the Display window of System Preferences.

- **Sound** offers a volume slider for your speakers, as well as a ◉ icon that turns blue when you're playing sound through a wireless speaker, headphones, or earbuds. Click the tile to see another volume slider; the names of any microphones, speakers, and other sound gear you've connected to the Mac; and a link to the Sound window of System Preferences.

- **Now Playing** shows the name, performer, and album art for whatever music is playing at the moment. It also offers a tiny play/pause button (▶︎ⅠⅠ) and a fast-forward/next track button (▶▶). Click the tile to see a scroll bar for the song, so you can jump around in it, as well as a full set of ◀◀, ▶, and ▶▶ buttons.

- **Accessibility Shortcuts.** This tile turns blue if you've activated one of the Mac's assistive features, like **Zoom** (enlarge part of the screen) or **Invert Colors** (view the screen like a film negative, to make things easier to see). Click the tile to see a list of the accessibility features (page 397), which you can click to turn them on or off. There are also links to the Keyboard Shortcuts pane in System Preferences, so you can set up keyboard short- cuts to turn features on or off, and to the Accessibility window of System Preferences.

- **Battery.** On a laptop, this tile shows how full your battery is, both as a graphic gauge and as a percentage. Click the tile to see the status of your laptop's power situation, such as **Power Source: Power Adapter** or **Fully Charged**; a list of any apps that are using an alarming amount of power; and a link to the Battery window of System Preferences.

- **Fast User Switching.** As you can read in Chapter 13, different people can use the same Mac at different times. When your sister, for example, signs in with her name and password, she sees her own files, folders, email, web history, and so on. Click the tile to see the icons of everybody with an account on this Mac. Your sister (or whoever) can sign into her account just by clicking her own icon. You also get links to the **Login Window** (another opportunity to change accounts) and the **Users & Groups** window of System Preferences.

> **NOTE:** This is the only Control Center tile that reveals nothing at all until you click it.

At the bottom of the Dock & Menu Bar window in System Preferences, there's a set of four final controls. As the heading in the list suggests, you can install these on the menu bar only:

- **Clock.** To the certain bafflement of millions, Apple moved the settings that govern your menu-bar clock *here* from the Date & Time preferences, where they've lived for generations. See page 54.

- **Spotlight.** Whenever you want to *find* something—on the Mac or on the web—you can click this Q and proceed as described on page 121. (You can also begin a search by pressing ⌘-space bar—and once you've learned that keystroke, there's no reason to clutter up your menu bar with the Q.)

- **Siri.** The voice assistant that iPhone made famous is available on the Mac, too (page 371). You can begin a spoken request by clicking this menu bar icon—but if you get used to saying "Hey Siri" instead, you can turn this icon off to save a little space.

- **Time Machine** is the Mac's automatic backup feature (page 64). If you don't use it, there's no reason to have its icon on your menu bar.

Set Your Notification Tolerance

Somewhere along the line, it occurred to some software engineer that a computer would be more useful if it took a more active role in getting your attention. You wouldn't have to check your calendar all day—the machine could display a reminder when you had a meeting or phone call coming up. You wouldn't have to keep checking your email—the machine could alert you with a chime when new email arrived.

These are *notifications*: alerts, messages, and warnings. They're a huge deal on smartphones—they call your attention to new text messages, appointments, Facebook or Twitter messages, low-battery warnings, and so

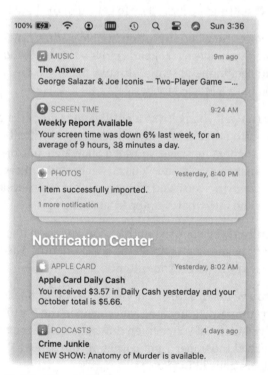

Notifications in Big Sur

on—but they're on the Mac as well, and in Big Sur they've been attractively overhauled.

A notification takes the form of a little translucent bubble at the upper-right corner of your screen.

Here's what you can do when you see one:

- **Read it and ignore it.** The Mac can display two kinds of notifications: *banners*, which appear briefly and then disappear (great for things like incoming email), and *alerts*, which stay on the screen until you click a button to get rid of them (good for alarms and flight updates). Neither interrupts your work. You can keep on doing whatever you've been doing.

> **TIP:** In Big Sur, you can see a little more context on a notification before you move on—at least in some apps, including Apple Podcasts, Mail, and Calendar. If you hold your mouse down on it, you get a pop-up panel that offers the opportunity to listen to the latest podcast, for example, or expand a Calendar invitation to see where it appears on the timeline of the day.

- **Dismiss them manually.** When you move your mouse onto one of these bubbles, a ⊗ button appears. Click to get rid of the bubble.

  You can also get rid of a notification bubble by dragging it off the screen to the right. If you have a trackpad, here's a third method: Point to a bubble (without clicking) and swipe with two fingers to the right—or with one finger if you have a Magic Mouse.

- **Respond briefly.** If you point to a notification without clicking, you reveal an **Options** button that offers ways to act on the notification. For example, for an incoming text message, you get a **Show More** button that lets you reply. If it's an incoming email, you can **Reply, Mark as Read**, or **Delete**. All this is very handy, because you can process the new development without having to switch into whatever app is speaking up.

- **Respond fully.** If you click the notification, you open its program, where you can read the message, peruse the article, or otherwise take care of whatever was trying to get your attention.

Just because you've taken care of a notification bubble doesn't mean it's gone forever, by the way. At any time, you can open the Notification Center—a column at the right side of your screen that shows all recent notifications—by clicking the clock at the upper-right corner of your screen. (You can see it in the figure "Notifications in Big Sur.")

Customizing Your Notifications

This, of course, is a chapter about setting up the Mac to make it the most useful and the least annoying. A big part of that job is controlling *which* notifications pop up on your screen. If you have 65 apps, and each one notifies you about every little thing, you'll lose your mind.

Open **System Preferences→Notifications**. Here you can individually tailor the notification behavior of each app. When you click an app's name, you get to make changes like this:

- **Allow Notifications** is the master on/off switch for this app. If you turn it off, you'll never be interrupted by notifications from this program.

- **Banners or alerts.** If you've permitted notifications from this app, you can choose which notification style you prefer—either **Banners** (which vanish after a moment) or **Alerts** (which stay on the screen until you click to dismiss them).

- **Show notifications on lock screen.** Do you want this app's notifications to appear on the screen even when the Mac is locked and asleep?

- **Show in Notification Center.** Do you want this app's notifications to appear in the Notification Center panel?

- **Badge app icon.** Do you want to see a little red numeric counter (❷) on this app's desktop or Dock icon, showing how *many* notifications have piled up?

- **Play sound for notifications.** When this app needs your attention, do you want it to play a little sound as the bubble appears?

- **Show previews.** Do you want the *contents* of the notification to appear in the bubble—the actual text/email/Facebook message, the details of the Uber pickup, the name of your alarm or to-do item? If there's any risk that some family member or co-worker might stroll by and read something embarrassing or incriminating, then maybe not.

    For each app, you can choose either **always** show the preview, **never**, or only **when unlocked** (previews will appear only when you're sitting in front of the computer working; they won't appear on the lock screen).

> **TIP:** You get this kind of **Show previews** control for each individual app, but at the bottom of the System Preferences window, you also have a master **Show previews** control that governs all of them at once. The app-by-app setting overrides this global one.

- **Notification grouping.** To save vertical screen space, the Mac can collect all the notifications from a particular app into a cluster, which expands when you click it.

In Big Sur, you can turn that clumping behavior **off**, on a per-app basis. Or you can choose **by app** to make each app's notification bubbles clump together. Or you can choose **automatic** from this pop-up menu. That setting groups *threads* together—that is, notifications about email conversations on the same topic.

Click to expand the piled-up notifications.   *Click to dismiss them all at once.*

Notification groups

You're not expected to get all your notification settings right on the day you first meet Big Sur. You *are* expected to remember that you can shut up the notifications from individual apps that become annoying.

Do Not Disturb

A "Do Not Disturb" sign, of course, is the classic hotel-room doorknob hanger that you put out when you don't want anyone to barge in. On the Mac, it's exactly the same idea: You don't want any notifications to appear, chime, or interrupt your work. Do Not Disturb is a master switch: It shuts up all interruptions, regardless of whatever individual app-notification settings you've established.

To turn on Do Not Disturb, open the Control Center (click 😎 on your menu bar), click **Do Not Disturb**, and then specify how long you want to be undisturbed. Your choices are **For 1 Hour**, **Until Tomorrow**, or **Always On** (which actually means "Until I turn you off again").

TIP: Super-quick way to turn on Do Not Disturb: *Option-click your menu-bar clock.* Option-click again to turn it off again.

System Preferences offers scheduling, exceptions, and other Do Not Disturb options.

The Control Center offers quick settings.

Do Not Disturb settings

If, at this point, you click **Notifications Preferences**, you'll see that you can set Do Not Disturb to invoke itself automatically according to a schedule—from 11 p.m. to 8 a.m., for example; whenever the screen or the Mac itself has gone to sleep; or **When mirroring to TVs and projectors**.

If you don't understand the purpose of that last option, then you've never experienced the appearance of a spouse's text message when you're onstage giving a talk in front of a thousand colleagues ("Hi Pookie! Don't forget to pick up your antifungal cream!").

Arrange Your Widgets

Not everything requires the creation of a full-blown app with its own name, icon, and associated bulk and complexity. That's why, in Big Sur, Apple has

Click to make your notifications and widgets appear.

Widgets in macOS

introduced *widgets*: compact floating windows that appear all at once, with a single click, displaying information like the stocks, the weather, your to-do list, and your upcoming appointments.

To see your widgets, click your menu bar clock (upper-right corner). Yes, that's also the technique for making the Notification Center appear; the widgets appear *beneath* the notifications.

You won't get much value from widgets until you've tailored which ones appear, and in which order, to serve you best. To do that, click **Edit Widgets** at the very bottom of the widgets display (you may have to scroll down to see this button).

At this point, your screen fills with a catalog of widget options. Just click one to install it into the right-side widgets panel.

While you're at it, you can rearrange your installed widgets by dragging them, or remove the ones you're not using at the moment by clicking the −.

> **TIP:** There's nothing to stop you from adding the same widget to your collection more than once. You might, for example, install one copy of the Weather or Clock widget for each city you want to track.

Available widgets...in a range of sizes

Installed widgets
(click to remove)

Editing your widgets

Software companies other than Apple can offer their own widgets; they're in the App Store like any other apps (page 131). But Apple starts you off with a selection of 12 widgets.

Most come in three sizes—small, medium, and large, which you can sample by clicking the **S**, **M**, or **L** button beneath each one. Bigger ones show more information but, of course, eat up more of your screen.

Here's a rundown:

- **Up Next** shows a slice of your calendar, so you can inspect your upcoming appointments.

- **City** is a clock that displays the current time. To specify the city, see "Tools on the Back of the Widget," below.

- **World Clock** shows the current time in *four* cities simultaneously. You specify which cities you want as described in "Tools on the Back of the Widget."

> **TIP:** On the back of the widget, the little ☰ handles let you drag the four chosen cities up and down to change their order.

- **Folder.** Once you get to know the Mac's built-in Notes app (page 274), you may come to adore it. You may even come to realize that you can create different "folders" full of notes on different topics. This widget displays the contents of one specified folder—whichever you choose (as described in the box below).

TOOLS ON THE BACK OF THE WIDGET

Many of macOS's widgets are designed to display a localized bit of information: the clock from a particular city, the contents of a particular note or Reminders list, the weather for a particular place, the price of a particular stock.

To specify which one you want a widget to show, scroll to the bottom of the widgets and click **Edit Widgets** (if you're not already in widget-editing mode).

Now click the face of the widget you'd like to fine-tune. It flips around with a cute little animation so you're looking at the "back."

When you now click the name of the city, folder, note, or whatever it is you're trying to specify, you can view a list of your options. The search box at the top helps you find what you're looking for.

With your selection specified, click **Done**. The widget flips around to the front again, with your new selection in place. ★

- **Note** lets you install, as a widget, a single note from your Notes app—something you might want for quick reference. Great uses for this idea: frequent phone numbers, a Brainstorms note, an Account Logins list, a packing list you've created in the app's checklist style (page 275).

  Once you've installed this widget, you can specify which note it displays according to the steps described in "Tools on the Back of the Widget."

- **Photos.** Awwww, what a sweet idea: a little floating panel that brings you a blast of sentiment—in the form of a Memory slideshow (page 283) whenever you need it. Each day, you'll see a different photo from your Photos collection; click it to open the Photos app, where you can click ▸ to view a slideshow of that Memory.

- **Up Next (Podcasts).** If you've been using Apple's Podcasts app, you can use this widget to pick up where you left off in one of your shows. The small version of the widget shows the most recent episode you've played; the larger ones show the last *few* episodes. Clicking either opens the Podcasts app itself so you can begin listening.

- **Reminders** displays the single most urgent To Do item in your Reminders app (page 302)—or, in the medium and large versions, a longer list of them. Click the widget to open the Reminders app itself to check them out—or check them off.

- **Daily Activity** gives you a quick and horrifying glance at how much time you've wasted sitting in front of your Mac today—a feature Apple calls Screen Time (page 392).

- **Symbol** shows you the current price and trend line of a particular stock. You specify it as described in "Tools on the Back of the Widget."

- **Watchlist** is also a stock ticker. In this case, though, you get to see the current share prices of three stocks on the small or medium widget, or six on the large one. This time, you can't choose these stocks on the "back" of the widget. Instead, the widget displays the contents of the "watchlist" you've created in the separate Stocks *app* (page 307).

- **Forecast** is, of course, a weather tile. The small size shows today's current and high temperatures; medium adds a five-day forecast; and large shows a five-day forecast *and* hourly conditions for today. To specify which city's weather appears here, see "Tools on the Back of the Widget."

Set up the Widgets display just right, and you'll discover that Apple has made these things useful and friendly. They save you a lot of web searches and app-opening when all you need is a quick lookup.

Design the Clock

In macOS Big Sur, you can always see the time, and it's always at the far right end of the menu bar.

The menu-bar clock options have moved to a new location:
System Preferences→Dock & Menu Bar.

You still set the time and time zone in
System Preferences→Date & Time.

Set up the time—and the clock

But in **System Preferences→Date & Time**, you can set that clock and specify exactly what you want it to look like.

Set the Time

To set the time and date, click the **Date & Time** tab. Now click the 🔒 and enter your Mac password (page 316). (Why is that security step necessary? Presumably so that naughty youngsters can't mess with your computer by changing its clock behind your back.)

Now you can tell the Mac to **Set the time and date automatically** (it consults the internet), or turn off that checkbox and set the time and date yourself. Use the little up/down arrow buttons, click the little calendar, or physically drag the hands of the clock.

Design the Clock

In pre–Big Sur days, the **Date & Time** panel also gave you some control over what the menu bar clock looks like: digital or analog, with or without the seconds, with or without the a.m. or p.m., with or without the day or today's date, and even as a military-style 24-hour clock (like "1830" instead of "6:30 p.m.").

In Big Sur, Apple moved all that stuff to **System Preferences→Dock & Menu Bar.** Scroll down the left-side list until you see **Clock.** Here are all the old clock options—including **Announce the time.** It makes the Mac mark the passage of your life by speaking the current time every hour, half-hour, or 15 minutes. (Just have pity on any co-workers who'll also have to hear it all day long.)

Auto-Type Your Email Address

It's one of the greatest features on the Mac, and maybe 2% of the world uses it: auto-typing. You set up short bits of text that, when typed, magically expand into much longer phrases, saving you time and eliminating typos.

The classic use of this feature is setting up @@ so it types out your complete email address, which is especially useful if yours happens to look something like *augustine.anastasia@millenniumcorp.com.*

But you might also set up the code *addr* to expand into your complete name, address, and phone number. Or maybe you do a lot of email work, and you find yourself typing the same phrases over and over. Set up one of these self-expanding abbreviations, and your fingers won't evolve into hideous claws from repetitive keyboard-pounding.

To create one of these expanding abbreviations, open **System Preferences**→
Keyboard. Click the **Text** tab and then the + button.

Now it's easy: Type your abbreviation into the left column, and type or paste
the expanded phrase you want into the right column.

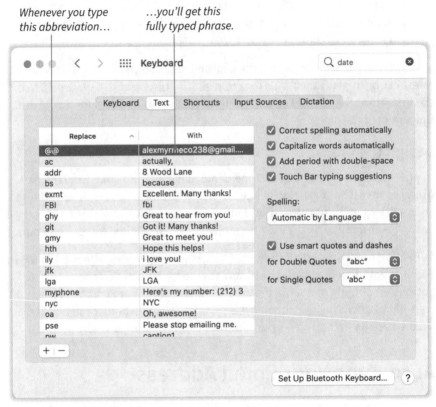

Self-expanding abbreviations

Once you've made up an auto-typed phrase or two, close System Preferences.
Now try it out: These abbreviations work in any window, in any app, where
typing is permitted.

Get Your Languages Ready

In theory, your Mac already speaks your language—the menus and apps
display their text in whatever language you chose the first time you set up
your Mac.

But in System Preferences, you can change the Mac's language—or even get it ready to switch among languages as you type. That's useful if you occasionally use other languages in your writing.

To add a second language, open **System Preferences→Language & Region**. Click + and choose from the long list of lovely languages. Your Mac can even handle languages that don't use the 26-letter Roman alphabet, like Japanese and Chinese, and languages you read right to left, like Hebrew and Arabic.

Each time you install a new language, the Mac invites you to make it your *primary* language—the main language for the entire Mac, for all your menus, buttons, and apps. If it's a language that doesn't use the 26-letter alphabet, you'll also be invited to choose an *input method*—a keyboard, palette, or trackpad drawing feature that lets you type or write the characters.

Click + to install a new language.

Then you can choose its name from the Input menu, even in midsentence.

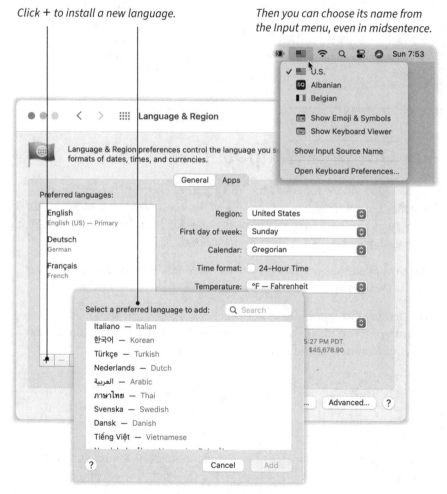

Installing new languages

If you do opt for a second input method, a new menu icon appears, indicating the current input method. In the U.S., this menu-bar icon starts out as a tiny American flag.

Now then. Suppose you're typing along, and you suddenly feel that *je ne sais quoi*, that feeling of *ennui*, that sense that English-only documents are your *bête noire*. Just choose the new language's name from the Input menu. (Its icon changes as you switch languages.)

Apple calls this little menu the Input menu—not the Language menu—for a good reason. Instead of listing languages, it lists ways of entering them. Sometimes that's a different layout of the keyboard (for example, pressing the ; key when you're typing in Swedish produces the ö symbol). Sometimes that's a little drawing pad so you can make Chinese characters with your finger.

> **NOTE:** If you don't see the little flag icon on your menu bar, open **System Preferences→Language & Region**. Turn on **Show input in menu bar** to make it appear.

In any case, now you're set to use your Mac in any language.

Pick Out Your Wallpaper

You paid good money for your Mac, and a decent chunk of it paid for your glorious, vivid color screen. You may as well show it off by choosing a great picture to use as its *wallpaper*. That's the desktop background—the tablecloth underneath all your files, folders, and windows.

To look over your options, open **System Preferences→Desktop & Screen Saver**. Click the **Desktop** tab.

In the **Desktop Pictures** folder at left: a set of gorgeous full-screen images for use as your wallpaper. You'll find nature photos, artsy pictures, abstract swirls, fabric patterns, and solid colors.

> **TIP:** The images called Dynamic Desktop are especially cool. These are scenes whose *lighting changes* throughout the day, according to the way the sun would be hitting it in the real world.

You're also welcome to use one of your own photos. The **Photos** category shows all the pictures you've ever stored in the Photos app (page 282),

Apple's photography—or your own—can become your desktop wallpaper.

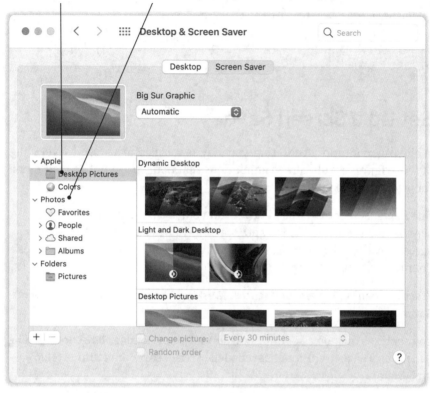

Designing your wallpaper

organized by the faces of the people in them or the albums that contain them. You can also click the + button under the list to choose any other folder of pictures on your Mac.

> **NOTE:** This may be your first encounter with *flippy arrows*—the Mac's system of displaying collapsing folders. When the **Photos** category is collapsed, for example, you're seeing only its name; the flippy arrow next to its name looks like this: ❯. But when you click, it rotates 90 degrees (❯) and shows the list of everything inside.

Once you've chosen a photo, by the way, a new pop-up menu appears at the top of the System Preferences window. It offers options like **Stretch to Fill Screen**, **Center**, and **Tile**. These are various ways of making your chosen wallpaper photo fit your screen, in case it wasn't born with precisely the same size and shape. Experiment to find something that doesn't look distorted and hideous.

Set Up a Screen Saver

While you're interior decorating, you may as well set up a screen saver—one of those fantastic full-screen animations that bounce around on your monitor when you're not using it.

The name "screen saver" is left over from olden times; today's flat-panel screens can't "burn in," so they don't actually need saving. Still, the Mac's screen savers look stunning, so you may as well choose one that enhances your reputation.

Start by opening **System Preferences→Desktop & Screen Saver**. Click the **Screen Saver** tab.

Your options include a set of 12 slideshow styles (**Floating, Reflections, Sliding Panels**) as well as traditional psychedelic screen saver animations (**Flurry, Arabesque**).

If you choose one of the slideshows, you then have to tell the Mac what pictures to use. Use the **Source** pop-up menu to choose a collection of **Landscapes, Flowers, Colors**, or whatever, or hit **Choose Folder** to pick out a folder full of your own photos (highly recommended, as long as they're safe for work).

For a quick demo of what the screen saver will look like, click the preview image. To exit the screen saver, click anywhere or press any key.

Don't forget to specify when you want this screen saver to come on. Use the **Start after** pop-up menu to indicate how many minutes of inactivity need to pass before the show begins.

You can also summon the screen saver instantly when it pleases you—by moving your mouse pointer into one of the four corners of the screen. To set that up, click **Hot Corners**. The four pop-up menus here represent the four corners of the screen. Use the appropriate menu to choose **Start Screen Saver**, and click OK.

Try Dark Mode

In dark mode, the backgrounds of your windows turn dark gray, and the text is white. Here's a comparison:

Dark mode   *Light mode*

Dark mode vs. light mode

Dark mode doesn't make you more productive. It doesn't save any battery power. *Maybe* it's less disruptive for somebody trying to sleep next to you in bed. Mainly it just looks cool.

To try it out, open **System Preferences→General**. At the top, click **Light**, **Dark**, or **Auto**—which uses light mode during the day and dark mode at night.

You'll quickly discover that dark mode mostly affects the Mac's built-in programs, like the Finder, Calendar, Photos, Messages, Notes, and Mail. Dark mode may not affect apps from other software companies at all.

Furthermore, dark mode usually doesn't affect what's *in* your apps' windows—the text you're writing, the photo you're editing, the numbers you're crunching. Mostly, it affects the window pieces *around* what you're working on: the menus, scroll bars, backgrounds, toolbars, palettes, and so on.

Give it a try. If you hate the look, restoring light mode is only one click away.

Turn On Insta-Zoom

There are all kinds of reasons you might want to zoom in on a particular spot on your screen. Magnifying tiny type is one obvious example, but sometimes you'll see somebody's tiny headshot icon on social media and want a closer look. Or maybe you're looking over a scan of a document, and you're having trouble making out a couple of the words.

The Mac has a fantastic feature that lets you magnify the spot at the cursor, to exactly the level you find useful. But until you turn it on, that feature lies dormant. Here's how you get it ready.

Open **System Preferences→Accessibility**. Click **Zoom**.

- **If you have a trackpad, a Magic Mouse, or a mouse with a wheel on it:** Turn on **Use scroll gesture modifier keys to zoom**. From now on, whenever you "scroll upward" while pressing the Control key, you zoom in, magnifying the screen.

  For example, the way you scroll up with the trackpad is to drag upward with two fingers. If your mouse has a wheel on top, you roll it away from you.

  If you "scroll down," still pressing the Control key, you zoom out again.

- **Otherwise:** Turn on **Use keyboard shortcuts to zoom**. From now on, you can enlarge the screen by 200% by pressing Option-⌘-8. At this point, you can adjust the amount of magnification by pressing Option-⌘-plus or Option-⌘-minus (the + and - keys).

  Press Option-⌘-8 again to return everything to normal size.

Fool around with this feature a little bit, and then don't forget you have it available. It's a slick trick that comes in handy.

Adjust Screen Locking

You might keep private stuff on your Mac: medical records, finances, a diary of your innermost thoughts, your plans to take over the world. That's why your account is protected by your password or fingerprint.

But, in theory, anytime you step away from your Mac, some passing evildoer could start snooping around. That's why it might be smart to set up auto-lock, where your Mac locks itself after a few minutes of activity, and your password (or fingerprint) is required to get back in.

> **NOTE:** Almost all Macs go to sleep, or display a screen saver, after a few minutes of inactivity; to resume what you're doing, all you have to do is click once or press a key. (Read on to specify how soon it goes to sleep, or see page 60 for how quickly the screen saver appears.) What you're doing here is requiring your password or fingerprint instead of just a key press to get back in.

To set this up, open **System Preferences→Security & Privacy**. Click the **General** tab. Turn on **Require password after sleep or screen saver begins**. From the pop-up menu, you can specify how quickly the Mac will lock the screen after the sleep or screen saver has kicked in.

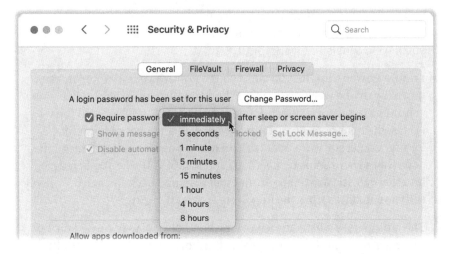

Setting up screen locking

Once you've added this layer of security, you'll feel much more comfortable leaving your desk for bathroom breaks.

Set Up Auto-Sleep

As you may recall from Chapter 1, you're not supposed to shut your computer down completely when you're done for the day. Instead, you put it to sleep, which saves you time. (Starting it up from scratch takes over a minute. Waking it takes about a second.)

As you may also recall, you can set up the Mac to go to sleep automatically if it notices that you haven't been doing anything for a while. To specify when that happens, open **System Preferences→Battery→Power Adapter** (laptops) or **System Preferences→Energy Saver** and adjust the slider called **Turn display off after** to specify a sleep interval after your last mouse or keyboard activity. (On Macs made after 2013, the rest of the computer—the processor, drive, and so on—sleep promptly and automatically.)

> **NOTE:** On a laptop, you also get a **Battery** tab with its own **Turn display off after** slider. That's because you may want to set up different auto-sleep intervals. You might want the Mac to sleep sooner when it's running on battery power, to save juice, but take longer to go dark when it's plugged in.

And while you're fiddling with your Mac's power settings, don't miss these options:

- **Schedule.** These controls make the Mac start up and shut down (or sleep and wake) automatically, at the same times every day. That's convenient if you have a consistent work routine.

- **Enable Power Nap.** Just because you're not using your laptop doesn't mean it can't be working for you. When you turn on Power Nap, the Mac continues to perform network tasks even while it's asleep and the lid is closed. About once every hour, it wakes up and checks the internet for new email, software updates, calendar updates, and updates to the apps that synchronize using iCloud (page 150). If the laptop is plugged into the wall, it can even perform Time Machine backups and download software updates from the App Store (page 131).

  The laptop uses a little more power when Power Nap is turned on. But waking it up to find that everything has been synchronized with the internet while you were away is well worth the trade-off.

Create Backups

Of all the misfortunes that are within your control to prevent, very few are as painful as losing your hard drive. All your photos. All your home movies. All your work files. All your personal files. Gone.

Truth is, you don't encounter as many sobbing lost-data victims these days as you once did. In part, that's because so much of our lives are now "in the cloud"—that is, stored online. You hardly ever hear of people losing all their email, for example. Even if your Mac dies, all your email is still there on Gmail, Yahoo Mail, Live Mail, AOL Mail, or whatever mail service you use.

Hardly anyone loses their music collection to a hard drive crash, because most of the world now listens to music that streams from services like Apple Music or Spotify.

If you use iCloud (Chapter 7), then your photos, notes, reminders, address book, passwords, text messages, bookmarks, audio recordings, and documents can be backed up online, too.

But backing up so much stuff on iCloud costs money, and it still doesn't include safety copies of your apps, all their settings, and other important elements of your digital life. For that reason, it's an excellent idea to set up a system that backs up your entire Mac automatically.

Fortunately, macOS comes with exactly such a feature. It's called Time Machine. You set it up once and never have to think about it again. It creates a constantly updated safety copy of your entire Mac. If disaster ever strikes, you'll be so happy you made the effort.

Setting Up Time Machine

Time Machine requires a second hard drive for its backup. For most people, that's an external USB hard drive; a 2-terabyte model, which is capacious enough to back up most Macs, costs under $65.

Time Machine backs up your entire drive, unless you click Options and choose some folders to leave out.

Time Machine preferences

The first time you connect such a drive, the Mac asks if you want to use it as your backup drive ("Do you want to use 'Western Digital 2TB Drive' to back up with Time Machine?"). Click **Use as Backup Disk** to open the Time Machine pane of System Preferences.

After a couple of minutes, the Mac begins the process of copying everything on your drive onto the Time Machine backup disk. Beware: This process can take a *very* long time—a couple of days, in fact, if there's a lot of stuff on your Mac.

Fortunately, the backup process is quiet and invisible and doesn't affect you. You can keep on using the Mac as you always do. (You can always peek at its progress—or pause it—using the commands in the ⊕ menu on your menu bar.)

But if Time Machine created a full duplicate of your entire Mac once per hour, the backup drive would soon overflow. So it works like this:

- **During the day,** Time Machine updates your backup once an hour. If a meteorite strikes your Mac when you're in the middle of writing your great American novel, the most you'll lose is 59 minutes' worth of work.

- **At the end of the day,** Time Machine deletes those hourly backups; it keeps only the last one for the day.

- **At the end of the month,** Time Machine preserves only one backup from each week.

In other words, Time Machine doesn't keep just *one* copy of your files; it memorizes what was on your Mac in such a way that you can rewind to any hour today, any day in the past week, or any week in the past month. It keeps on making new snapshots of your Mac's drive until the backup drive is full. At that point, it deletes the oldest backups to make room for new ones.

Recovering Files

Suppose one day you just can't find a certain file. Or you've *changed* a certain document in a way you now regret, and you want to rewind it to an earlier condition. That's when Time Machine swoops in to save you.

To start the rewinding process, open the ⊕ menu on your menu bar. (If you don't see it, open **System Preferences**→**Time Machine** and turn on **Show Time Machine in menu bar.**)

Now the screen changes. Suddenly, you see your regular Finder desktop—or, rather, dozens of copies of it, stretching away. Each is a snapshot of whatever was in that window at the time of a backup.

Use the arrows, or the time scale, to rewind this window to an earlier state, back when your lost files (or yet-unedited files) still existed.

Entering Time Machine

To explore these past versions of your Mac, you can use any of these techniques:

- **Explore the folders.** You can look through the contents of your files and folders manually, as described in Chapter 4.

- **Click the ⌃ button beside the window.** This arrow button jumps to the most recent version of the currently open window that's *different* from the way it is right now. That is, if the contents of this window last changed nine days ago, one click of that ⌃ button takes you to it.

- **Drag through the timeline.** The tick marks on the vertical "ruler" at the right edge of the screen represent individual backups. By dragging through them, you can jump directly to a certain date.

- **Click** Q to search for a certain file or folder within the backup you're looking at right now.

As you look through these backups, you can use some of the standard Finder navigation techniques described in Chapter 4. For example, you can use Quick Look (page 103) to preview what's in a file to make sure you've got the right document, or the right version of it. You can also change views—icon view, list view, and so on (page 96).

If your little time-travel experiment has successfully located a *deleted* file or folder, select it and then click **Restore** (at bottom). The Mac puts it right back in the window it started from.

If instead you're trying to recover an *older version* of a file or folder, select it and then click ∧; Time Machine skips to the most recent version that's different from the current one. Once you find it, you can use the **Restore** button to recover it.

> **TIP:** Time Machine also works to recover emails or address-book cards you've deleted. Once you're in Mail or Contacts, click the Time Machine icon (⏱) on the Dock. Once again, you enter the recovery mode—but this time, it looks like a strange, stripped-down copy of Contacts or Mail. You can navigate and recover a message or a Contact card using the same techniques you use to recover a file. (You can even use the search box in the Time Machine version of Mail or Contacts.) If you recover a deleted message this way, you'll find it in the **On My Mac→Time Machine→ Recovered** folder at the left side of the window.

Recovering the Whole Hard Drive

Now suppose your luck *really* runs out: Your Mac gets stolen, riddled with bullets, or thrown over the railing of your ship. Everything on it is gone.

Fortunately, Time Machine backs up your entire Mac. So once you have a new Mac or a working drive, Time Machine can rewind your computing life to a happier time. Begin by plugging your Time Machine backup disk into the Mac (with its USB or Thunderbolt jack, for example), and then proceed like this:

1. **Start up your Mac in recovery mode.**

   Page 368 describes the three ways to enter recovery mode. They all offer an option called **Restore from Time Machine backup.**

2. **Choose "Restore from a Time Machine backup." Click Continue.**

   Now you're shown any Time Machine backup disks the Mac knows about. You probably have only one.

3. **Click the name of your Time Machine backup disk, and then, in the list, click the most recent backup.**

   The installer starts copying your digital universe from the backup disk onto your new, empty hard drive. When it's all over, your Mac is exactly the way it was before disaster struck.

CHAPTER THREE

Laptop Life

ere's a statistic that may surprise you—or maybe not: 80% of all Macs sold are laptops. Mac fans may devour articles about the latest desktop models, like iMacs and Mac Pros, but what most people use are MacBooks.

It makes sense, actually. You don't pay much of a power or price penalty with a laptop, and you gain portability, which is a huge deal. Even if you don't work on airplane trays and hotel-room beds, you can still move the thing around your house, which is not nothing.

But in a few small ways, life with a laptop isn't the same as using a desktop computer, as the following pages make clear.

The Trackpad

The Apple mouse may have been a revolutionary pointing device when it came along in 1984, but it's not ideal for laptop use. What are you going to do, roll it across the leg of the guy next to you on the plane?

Of course, when you do have a desk, you can always plug the mouse into the laptop. The rest of the time, though, you're supposed to point and click using the trackpad.

The MacBook-family trackpad harbors a couple of secret features. If you never knew about them, your life would still go on—but they could bring you some time and efficiency.

Multitouch

The MacBook trackpad is a *multitouch* trackpad, meaning it can detect the touch of several fingers simultaneously. That ability unlocks some sneaky

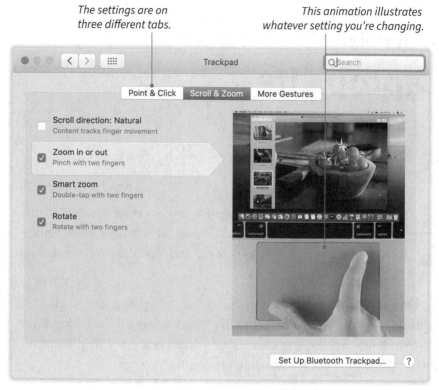

The settings are on three different tabs.

This animation illustrates whatever setting you're changing.

Trackpad settings

shortcut features, which you can explore in **System Preferences→Trackpad**. Here are the best of them, whose on/off switches are on the **Scroll & Zoom** and **More Gestures** tabs.

- **Zoom in or out.** When this feature is on, you can magnify things on the screen using your trackpad exactly as on an iPhone or iPad: Spread two fingers apart to magnify; pinch together to shrink it down. These gestures don't work in every app, but they're available in most software from Apple (Photos, Preview, Safari, TextEdit, iMovie, Pages, Keynote, Numbers, Final Cut Pro, Maps), Microsoft (Word, Excel, PowerPoint), Adobe (Photoshop, InDesign, Illustrator), and others.

TIP: In many of those programs, you can *rotate* the image by putting two fingers on the trackpad and twisting as though you were holding an invisible knob. Turn on the **Rotate** checkbox to make it possible.

- **Switch between web pages.** In a web browser, like Safari or Chrome, you can swipe across the trackpad with two fingers to move among web pages you've had open recently. Swiping to the left is the same as clicking the ‹ button; to the right is like clicking ›.

- **Notification Center.** If you swipe inward from beyond the right edge of the trackpad with two fingers, you summon the Notification Center and the widgets described in Chapter 2.

- **Show desktop.** When you splay your thumb and three fingers outward on the trackpad, all your open windows slide out to the edges of your screen, revealing the desktop underneath (page 83). That's handy when you want to dive back to the Finder to look for some file. Scrunch those four fingers inward again to bring everything back.

Force Touch

Mac laptops made since about 2015 have what Apple calls Force Touch trackpads, meaning they're capable of detecting how *hard* you're pressing down. This design, too, unlocks a few interesting features.

One example: Ordinary trackpads are hinged at the top and therefore require more pressure to click when your fingers are closer to the top. But on a Force Touch trackpad, you can click anywhere on it with the same amount of pressure—a pressure level, furthermore, that you can adjust in **System Preferences→Trackpad→Point & Click.**

Behind the scenes, a Force Touch trackpad doesn't actually move at all. It has a tiny vibration module underneath that makes it *sound and feel* like you've clicked it, but that's just a fakeout.

> **TIP:** How do you know if you have one of these newfangled trackpads? When the computer is turned off, the trackpad doesn't seem to click at all.

But you can actually *do* things with Force Touch, too. For example, you can "force-click," which means to click—and then, in mid-click, press harder to trigger a deeper click. That lets you do things like this:

- **Preview a document.** At the desktop, force-click any icon (like a picture, music, word processing, PDF, or movie file) to see what's inside it, without actually having to open its app. It's a shortcut for Quick Look (page 103).

- **Get the details.** In some programs, force-clicking opens a little window that reveals more information about whatever you clicked. It works that way in Messages (force-click a person's name in the conversation list), Reminders, and Calendar.

- **Look up a definition.** In Mail or Safari, force-click a highlighted word or a phrase to see more information about it from sources like Dictionary, Wikipedia, and more.

Force-click any word to see its definition.

In early March, Jimmy Chin, a professional photographer and mountaineer, finished back-to-back work trips — climbing in Antarctica, filming in Chile — and flew home to Wilson, Wyo. He was in the midst of plotting the rest of a typically peripatetic year, with trips to Thailand, New

Dictionary

per·i·pa·tet·ic | ˌperəpəˈtedik | adjective
1 traveling from place to place, in particular working or based in various places for relatively short periods: *the peripatetic nature of military* more

Thesaurus

peripatetic adjective
his peripatetic way of life: **NOMADIC**, itinerant, traveling, wandering, roving, roaming, migrant, migratory, ambulatory, unsettled, vagabond, more

, when the pandemic
in 20 years," Chin,
made do, editing a
running nearby
try and racing up and
775-foot peak that
st amateur climbers.

Pop-up definitions

- **Addresses.** In almost any Apple app—Mail, Contacts, Maps, Notes, Safari, and so on—force-click a highlighted address to open a map of its location.

- **Appointments.** Force-click a highlighted time and date—in any Apple app—to open a Calendar preview window, so you can see if you're free and even add the event to your schedule.

- **Link previews.** In Safari and Mail, force-click a web link to get a preview of the page that would open if you clicked it. It's like being able to see around the corner.

- **Renaming icons.** Force-click an icon's name in the Finder to open its renaming rectangle.

- **Annotating attachments.** When you're about to send a message containing a graphic or PDF attachment, force-click its icon to activate the markup tools (page 297).

- **Flight numbers.** In any Apple app, force-click an airline flight number (like *Delta 28*) to read details about the flight and the airplane.

- **Tracking numbers.** In Safari or Mail, force-click a UPS or FedEx tracking number to read details on where that package is.

Here and there, a Force Touch trackpad lets you *vary* the amount of pressure you're using to click. In QuickTime and iMovie, for example, when you're clicking the rewind or fast-forward buttons, pressing harder makes the movie scan faster. Same thing when you're pressing the arrow buttons in a Photos album: Press harder to flip through them faster. In Maps, press harder to zoom in faster.

There are a couple of downsides to force-clicking. First, because nobody was born knowing the list of places it works, it's a completely invisible and unused feature for most people. Second, lots of people wind up triggering a force-click accidentally. Some weird pop-up bubble appears without your understanding why, and you get frustrated.

If you're in that category, you'll be happy to know there's an on/off switch for this feature—in **System Preferences→Trackpad→Point & Click.**

The Touch Bar

On MacBook Pro laptops, Apple has replaced the traditional row of function keys at the top of the keyboard with a colorful illuminated strip about half an inch tall. It's the world's shortest touchscreen, and it's called the Touch Bar.

These buttons appear, disappear, or change depending on the app.

The Control Strip

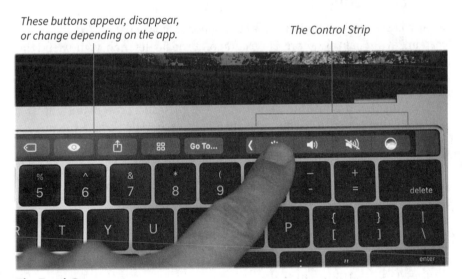

The Touch Bar

What it shows depends on what you're doing and what app you're using. It's supposed to give you quick access to commands that would otherwise require you to fiddle around in menus.

For example, when you're browsing the web in Safari, the Touch Bar displays tiny images of the different web pages you might have open, so you can hop among them. When you're typing, it displays autocomplete buttons for the three words the Mac thinks you're most likely to type next. When you're reading an electronic novel, it shows something like a map of the entire book, so you can zoom to any point with a finger touch.

> **NOTE:** If you don't use it for a while, the Touch Bar fades to black to save power. Just tap to make the buttons light up again.

And now the bad news: As cool as the Touch Bar concept must have sounded in Apple design meetings, in practice, many people don't use it for anything besides adjusting the Mac's brightness and volume.

> **TIP:** In a few programs, notably Microsoft ones, the traditional function keys (F1, F2, F3…) sometimes trigger valuable functions. Fortunately, having a Touch Bar doesn't mean you have to do without them. You can hold down the physical **fn** key on your keyboard to make the old F-keys appear on the Touch Bar.

The Control Strip

What you see on the Touch Bar changes depending on what you're doing, but certain keys are always there. For example, at the far-left end, your friend the Esc key is usually there, ready to close a dialog box, back out of a menu, or exit full-screen mode.

On the other hand, at the far-right end, you've got a short row of buttons called the Control Strip. It offers the absolutely essential buttons: screen brightness, speaker volume, and Siri the voice assistant.

When the Control Strip is in its space-saving collapsed form, you have three ways to adjust the brightness and volume:

- **Tap ☼ or ◀».** A slider appears. Drag your finger along it to adjust the Mac's screen brightness or speaker volume.

- **Slide out from the ☼ or ◀» button.** Keep your finger down after the initial tap, and drag within the slider.

- **Make little swipes right on the key.** If your laptop was made in 2018 or later, you can adjust the brightness or volume using an even more direct technique: Make short little horizontal swipes directly *on* the ☼ or ◄)) buttons. No slider ever appears, and you save a couple of seconds.

Of course, brightness and volume aren't the only useful top-row keys the Touch Bar replaces. What if you need the keys that open Mission Control, operate music playback, or change the keyboard illumination?

You can bring them back. Tap the thin ❨ button to the left of the ☼ button. The Control Strip now expands across the full length of the keyboard, revealing the same set of function keys that Touch Bar–deprived Macs have.

Tap to re-collapse the Control Strip.

The function keys return

> **TIP:** You can also set things up so the standard function keys are always visible on the Touch Bar. To do that, open **System Preferences→Keyboard**. From the **Touch Bar shows** pop-up menu, choose **Expanded Control Strip**.
>
> (That same pop-up menu offers a choice called **App Controls**. If you choose that option, the Control Strip is always hidden, leaving a lot more space on the Touch Bar for function buttons.)

Customizing the Touch Bar

In a few apps—including the Finder, Mail, Photos, and Safari—you can edit the Touch Bar: You can change what buttons appear there. The trick is to choose the **Customize Touch Bar** command, which is usually in the **View** menu. A bizarre dialog box appears on your screen.

Now you can proceed like this:

- **Delete a button** from the Touch Bar by dragging it with a finger all the way to the 🗑 icon at the left end.

- **Rearrange existing buttons** by dragging them around with your finger.

- **Install a new button** by dragging it downward from the laptop screen, using your mouse or trackpad, "through" the laptop's hinge, and onto the Touch Bar itself. If you drop it onto a spot marked by ---- dashes, you're *adding* it to the Touch Bar. If you drop it onto an existing button, you *replace* that button.

In each case, hit **Done** to wrap things up.

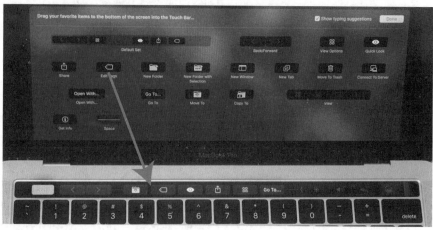

Customizing the Touch Bar

You can use a similar method to edit the Control Strip (the buttons always visible at the right end of the Touch Bar). Apple believes the functions you'll use most of the time are Brightness, Volume, Mute, and Siri, but you may not agree. Maybe, for example, you can do without the Siri key, since there are so many other ways to trigger Siri. Or maybe you'd like to add a button for Screen Saver, Show Desktop, or Sleep.

To make these changes, open **System Preferences→Keyboard**; click **Customize Control Strip**. At this point, you can edit the Control Strip buttons just as you would the main Touch Bar buttons.

The Fingerprint Reader

The best part of the Touch Bar may be the little gap, or the little black key, at the far right end. That's the Touch ID sensor—a fingerprint reader—that lets you log into your Mac, and sign into many websites, with just a finger touch. That's a huge improvement over typing your name and password 579 times a day. (Even some Mac laptops that don't have the Touch Bar *do* have the Touch ID reader.)

To put Touch ID to work, open **System Preferences→Touch ID**. Click **Add a fingerprint**, and follow the instructions to teach the Mac what your fingerprint looks like. (You can teach it up to five different fingerprints. Having a backup can be handy when you've got ketchup on your main finger.)

Once that's done, the next time you're facing the Mac's "Enter password" screen, just rest your finger on the Touch ID button for a moment to blow right on past it.

Your fingerprint can also serve as your key to making one-touch purchases on any website that offers an Apple Pay option.

No more entering your name, address, phone, and credit card number to check out from a web store. Your fingerprint unlocks all of it.

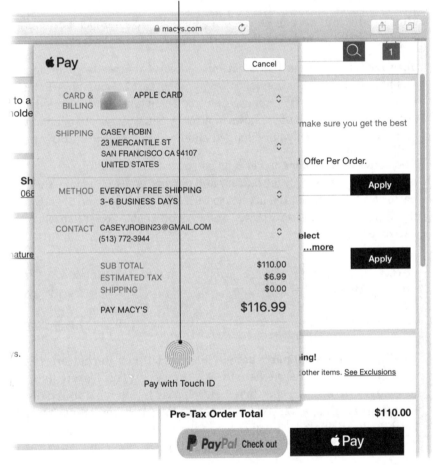

Apple Pay on the web

The Long-Lived Battery

Here's something you don't have to think about with a desktop Mac: running out of juice. But when your MacBook isn't plugged into the wall, it's running on battery power.

Apple may advertise "all-day battery life," but that comes with enough footnotes to fill a podiatry journal. How long your battery lasts depends on a lot of things:

- **What you're doing.** If all you're doing is reading websites, you really can go all day on a single battery charge. Playing video games? Not so much.

> **TIP:** For the longest battery life, quit any programs you're not using.

- **How bright your screen is.** The screen is a laptop's number-one energy consumer. The dimmer you keep it, the longer the MacBook runs.

- **Your Battery settings.** In **System Preferences→Battery→Usage History**, you'll find a big Big Sur bonus: battery graphs! **Last 24 Hours** shows the hour-by-hour level of your charge, plotted against how much time the screen was on. **Last 10 Days** shows how many batteries' worth of charge you used each day. (If you recharged midday, it might be over 100%.)

The **Battery** and **Power Adapter** tabs let you set up the laptop's energy appetite independently for battery and plugged-in use. To make the battery last longer, set the screen to turn off sooner, turn on **Slightly dim the display while on battery power**, and turn off **Enable Power Nap while on battery power.**

> **NOTE:** Recent laptops offer a strange new option here called **Optimize video streaming while on battery**. That's a reference to *HDR* movies (high dynamic range), meaning "great-looking color … which requires more power to display." Turn this on if you can do without the HDR glory when you're streaming video without a power cord.

Finally, don't miss **Optimized battery charging**. Lithium-ion batteries live shorter lives if you charge them to 100% every day. This feature limits charging to 80% *except* when it anticipates, by studying your daily charging patterns, when you might need a longer charge. (Hint: It won't be able to anticipate that you're flying across the country tomorrow. Tonight, turn **Optimized battery charging** off.)

> **TIP:** The ▬ menulet reveals whether you're running on battery or power adapter, estimates the time remaining on this charge, and identifies which apps are drinking down the most power right now. If you're trying to prolong this particular charge, quit those apps.

- **Wi-Fi.** Wi-Fi is basically a radio transceiver—and if you turn it off, by turning off the switch in the 🛜 menulet, you save a little bit of juice.

- **How old it is.** Any lithium-ion battery, like the one in your laptop, can be recharged only a finite number of times before it begins holding less charge. After you've recharged your MacBook 500 times, each charge lasts for only about 80% as long as it did when it was new.

> **NOTE:** In very cold weather, you may notice that a battery charge doesn't last as long. (It gets back to normal once you're at room temperature.) But time in very hot weather, over 95 degrees Fahrenheit, can *permanently* limit the amount of charge it can hold. Bottom line: The MacBook prefers temperatures between 50 and 95 degrees.

If you're not going to use the laptop for a while, store it with half a charge. If you store it dead, it may never be chargeable again, and if you store it fully charged, it may lose some charging capacity.

Eventually, if you use a MacBook for enough years, the Control Center's Battery tile may say **Service Recommended**. What it's trying to tell you—in case you haven't already noticed—is that your battery no longer holds a full charge, and it's time to consider visiting an Apple Store to get it replaced.

Personal Hotspot

Apple makes four kinds of devices intended to be mobile: watches, phones, tablets, and laptops. You can buy three of them with the built-in ability to get onto the internet over the cellular network—almost anywhere you go—instead of being limited to Wi-Fi hotspots.

But MacBooks are not so lucky. (What does Apple have against its laptops?)

On the iPhone, in Settings, confirm the password.

On the Mac, choose your phone's name. You're now online.

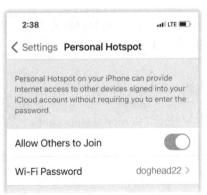

Personal Hotspot

At least there's a solution, called the *Personal Hotspot*. If you do have an iPhone or a cellular iPad, you can use it as a glorified wireless internet antenna for your laptop nearby. That way your MacBook feels as though it, too, has been blessed with the miracle of cellular internet.

Here's how to set it up. On the phone, open **System Preferences→Personal Hotspot**. (If you don't see it, open **System Preferences→Cellular** instead.) Make sure **Allow Others to Join** is turned on, and make a note of the password. Feel free to tap **Wi-Fi Password** to change it to something you prefer.

Now, on your Mac, open the 🛜 menu—the list of available Wi-Fi networks—and choose your iPhone's name.

Like magic, the 🛜 symbol on your menu bar changes to ⊚ ... a blue bar appears at the top of your iPhone to indicate that another device is mooching its internet connection ... and your laptop is now online, courtesy of the nearby phone or tablet. You probably never even had to enter the password.

> **NOTE:** That no-password business is a feature Apple calls Instant Hotspot, and it's part of the suite of iPhone/Mac wireless connection features called Continuity (page 152). Continuity requires that each device has Bluetooth turned on, has Wi-Fi turned on, and is signed into the same iCloud account (page 150). If any one of those things isn't true, then you *do* have to enter the iPhone's Personal Hotspot password at this point.
>
> The password is also necessary, of course, for anyone *else* who wants to use your phone's Personal Hotspot.

There is a price to all this magic, however. The first is battery power: Personal Hotspot drains your iPhone's charge right quick. Unless the phone is plugged into power, it's probably wise to save this feature for quick email checks or web articles, not streaming Netflix miniseries.

There may be a financial cost, too. Check your carrier's website to see how much Personal Hotspot use is included with your plan. (They may call it *tethering*.)

On the other hand, Apple's version of Personal Hotspot is especially nice. First, your iPhone's hotspot name always shows up on your Mac's 🛜 menu, ready to use, even if the phone is asleep, and even if Personal Hotspot is turned off. Second, if the phone or tablet is running iOS 13 or later, your Mac stays connected to the hotspot even if you close the lid or put it to sleep. That way it can still download new email and messages when you're not actively using it.

PART TWO

Welcome to Big Sur

CHAPTER FOUR

Home Base: The Finder

W hether it's a phone, a tablet, a laptop, or a desktop, every modern computer has some kind of starting screen. It needs a home base—a screen that displays all your files, folders, and apps. On the Mac, it's the desktop, which is also called the Finder.

Once you've logged into your Mac (page 316), the desktop is the first thing you see. On a brand-new Mac, its primary feature is a flowing rainbow of color stripes. That's not a bad choice of wallpaper, but of course you're welcome to replace it with any image you prefer (page 58).

Sidebar   A Finder window   Menulets

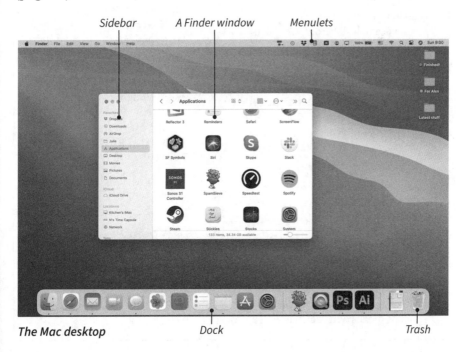

The Mac desktop   Dock   Trash

Some people prefer to keep the desktop tidy, unmarred by icon clutter, allowing their wallpaper to shine in its full glory. Others prefer to turn the desktop into a garden of choices, blooming with hundreds of files and folders that sit where they fell at the moment of creation, much in the same way some people use their physical desktops. Both approaches are legitimate (on the computer, anyway).

Menulets

As you know from page 3, the menus are the command lists that sit at the top of the Mac screen.

MENU KEYBOARD SHORTCUTS

Depending on your level of geekiness, you may come to wish there were a way to operate menus without having to move your hand to the mouse—an unnecessary expenditure of three calories. Often there is.

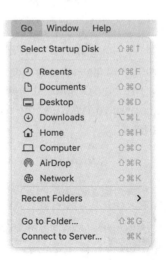

In a typical menu, some of the command names have symbols next to them—in Big Sur, they appear in gray. In most apps, for example, the **Bold** command in the menu is accompanied by the notation ⌘-B.

What it's telling you is this: Instead of using the mouse or trackpad to open a menu to choose the word **Bold**, you can just press ⌘-B. That is, while you're holding down the ⌘ key, type the letter *B* key, and then release both.

The Mac has several of these modifier keys, which, in combination, make possible hundreds of keyboard shortcuts.

To learn if a shortcut is available for a command, open the menu; the shortcut appears right next to the command name. What you might find confusing, though, is that most of the keyboard *symbols* in the menus don't actually appear

on your keyboard! What on earth does ⇧-⌃-⌥-⌘-W refer to?

In the menus, ⇧ refers to the Shift key, ⌃ is the Control key, ⌥ means the Option key, and ⌘ is the Command key.

If the command you want says ⇧-⌃-⌥-⌘-W, in other words, you're supposed to type *W* while pressing the Shift, Control, Option, and Command keys.

At that point, it might actually be quicker to use the mouse. ★

At the right end of the menu bar, though, you'll find a few menus represented as icons—which, because Apple hasn't named them, we'll call menulets. These don't change as you move among apps, either. They control essential Mac features like screen brightness and speaker volume. You get to choose which menulets appear here (page 41).

> **TIP:** If you rarely use the menu bar, or if you can't forgive the nearly half-inch of screen space it eats up, you can make it hide itself until you need it. Open **System Preferences→Dock & Menu Bar**. Turn on **Automatically hide and show the menu bar**. From now on, it's hidden until you summon it—by moving your mouse up to the top edge of the screen and waiting for half a second.

The Dock

A new Mac comes with 50 apps already installed—yours free!—and in time you may install many more. They all wind up in your Applications folder. (At the desktop, choose **Go→Applications** to see them.)

But, honestly, opening a folder containing 75 app icons is a pretty clumsy way to start your work each day. That's why Apple has given you the Dock: the floating strip of colorful app icons at the bottom of the desktop. They're supposed to represent the apps you use most often.

> **TIP:** Most things in the Finder require a double-click (two fast clicks) to open. Anything on the Dock, however, opens with just one click.

As a handy bonus, the Dock also tells you which programs are *open* at the moment, as indicated by tiny black dots beneath their icons. These apps are already in the Mac's memory, meaning you can return to them without the usual five-second wait.

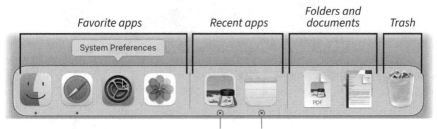

Parts of the Dock

Big Sur's redesigned Dock has three sections separated by vertical lines. First, there's a row of icons for your most frequently used apps, for quick opening. Next the Dock displays icons of apps you've used recently. On the far right, the Dock stores icons for things that aren't apps—like documents and folders—that you'd like to have available. The Trash is at the far right.

Apple makes a guess at which apps you might like to have installed on the Dock, but there's almost zero chance it was completely right. Fortunately, you can redesign the Dock like this:

- **Add a new icon to the Dock** by dragging it onto the appropriate section (apps on the left, documents and folders on the right). For example, you might open your Applications folder and then drag the icon of some new app directly down onto the Dock.

As you drag onto the Dock, the existing apps scoot aside.

Installing a new icon onto the Dock

- **Rearrange Dock icons** by dragging them horizontally.

- **Remove Dock icons** by dragging them up and away from the Dock, and then letting go of the mouse button.

By the way, the Dock doesn't have to take up space across the bottom of your screen. In fact, it may make more sense to have it take up space at the side of your screen, as a vertical column. Or maybe you'd just rather get rid of it.

Open **System Preferences→Dock & Menu Bar.** Here you have a slider that controls the size of the Dock's icons and a choice of positions on the screen. A couple of the other options here may be useful, too:

- **Magnification.** When you turn on this checkbox, each Dock icon swells to a bigger size as your mouse passes over it, making it a bigger target. Use the slider to specify how big.

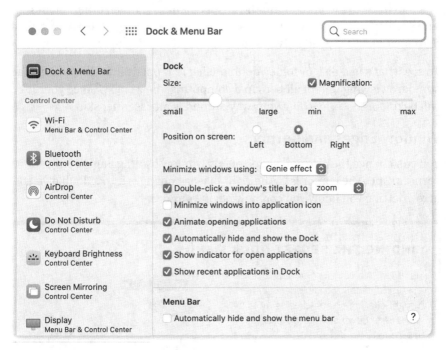

Dock & Menu Bar preferences

- **Automatically hide and show the Dock** means the Dock is ordinarily hidden, saving screen space and clutter. When you move your mouse to the corresponding edge of the screen, it reappears.

- **Show indicator for open applications** refers to the little dots beneath the icons of open apps. If you suffer from a rare condition that makes these dots remind you of bugs or dust, you can turn this off to hide them all.

How to Work Windows

When you get right down to it, the primary purpose of the Finder is to represent all your software as *icons* on the screen. Every app you use, every document you create, every disk and folder used to organize them—each one is represented by an icon (a little picture).

If you get even further right down to it, you can think of every icon as a *container*. A document icon might contain your writing, a spreadsheet, a photo, or a movie. Folder icons contain those documents, and even other folders. And you can put all of *that* onto disk icons, which represent hard drives, flash drives, cloud drives, and network drives.

To see what's in a disk or folder, double-click it; it opens into the rectangular standard viewing unit of all personal computing, the *window*. Since your life will be full of these, learning your way around is an essential skill.

Window Edges and Corners

You can change the size and shape of any window by dragging its edges or corners. Once the cursor is in the right place, it changes into a double-headed arrow, letting you know you can now begin your drag.

FINDING THE SECRET DOCK MENUS

Lurking within each icon on your Dock is a shortcut menu. You can open it either by right-clicking or long-clicking (holding your cursor down on the icon for a couple of seconds).

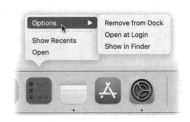

The commands inside include **Quit** (close the app if it's open), **Hide** (hide all the app's open windows), and **Options**. That last command is a submenu, listing three options of its own.

One of them, **Open at Login**, is particularly useful. It ensures that every time you sign into your Mac, that app will be open and waiting. It's a time-saver if you spend most of your time every day with certain apps—calendar, chat program, and web browser, for example. ★

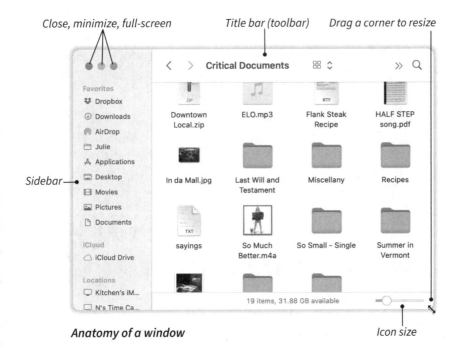

Close, minimize, full-screen   Title bar (toolbar)   Drag a corner to resize

Sidebar

Anatomy of a window   *Icon size*

The Title Bar

In most apps, including the Finder, the top edge of the window contains a horizontal bar. It always contains the three window-sizing buttons described next (red, yellow, green); it sometimes contains a toolbar or other controls; and in the Finder and many other apps, it displays the name of the window itself.

Most people probably imagine that the function of the title bar is to show you the name of the window, and that's often true. But the title bar also harbors a few secrets that can come in handy:

- **Move the window.** In most apps, including the Finder, you can use the title bar as a handle to move the entire window on the screen. Drag from any blank spot, or on the title of the window itself.

- **Enlarge the window.** In the Finder, double-clicking the title bar makes a window just big enough to show you all the icons inside it—or as many as it can fit without scrolling. In your other apps, double-clicking the title bar generally makes the window just big enough to reveal all the text, graphics, or whatever the window contains.

- **Identify the window in front.** Suppose you have several folder windows open, overlapping. The title bar helps you identify which one is in *front*—because all the background windows have dimmed, light-gray title bars.

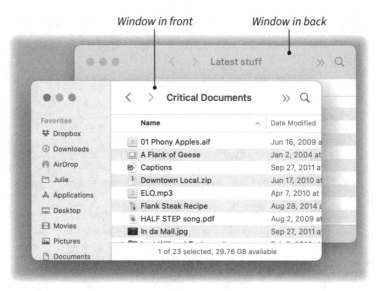

Window in front   Window in back

| | | Latest stuff | >> Q |

		Critical Documents	>> Q
Favorites	**Name** ^	**Date Modified**	
☘ Dropbox			
⬇ Downloads	🎵 01 Phony Apples.aif	Jun 16, 2009 a	
◉ AirDrop	🖥 A Flank of Geese	Jan 2, 2004 at	
🗁 Julie	▦ Captions	Sep 27, 2011 a	
⚘ Applications	❗ Downtown Local.zip	Jun 17, 2010 a	
▭ Desktop	🎵 ELO.mp3	Apr 7, 2010 at	
⊟ Movies	▦ Flank Steak Recipe	Aug 28, 2014 a	
🖼 Pictures	▪ HALF STEP song.pdf	Aug 2, 2009 a	
🗋 Documents	▦ In da Mall.jpg	Sep 27, 2011 a	

1 of 23 selected, 29.76 GB available

Foreground and background windows

Window-Sizing Buttons

At the top left of every window on the Mac—not just in the Finder—are three famous buttons, designed like a sideways traffic light: red, yellow, and green. As the tip of your arrow cursor passes over them, they sprout little symbols to hint at their functions:

- **Close (⊗).** Clicking the little red button in the top-left corner makes a Mac window go away. That's a really good one to remember.

> **TIP:** You can also learn the keyboard shortcut ⌘-W—for *window*, get it? That's especially handy if you're trying to close a bunch of windows, because you can just tap ⌘-W repeatedly without having to aim for successive Close buttons in different positions on the screen.

- **Minimize (⊖).** Click this button to get the window out of your way temporarily. It shrinks down into an icon at the right end of your Dock. It's not gone—just set aside until you click that Dock icon to bring the window back. Minimizing a window is a great way to see what was behind it, or to get it out of your hair while the window is finishing some task that you don't need to babysit.

- **Full-screen (⊘).** In most Mac apps, this button makes the window fill the entire screen, edge to edge. The point is to give you the largest possible canvas for whatever work you're trying to do (or video you're trying to watch, or game you're trying to play). To make that possible, just about everything except your work is *hidden*, including the menus, the Dock, scroll bars, status bars, tool palettes, and so on.

The menus and the Dock aren't completely gone, however—they're just hiding. Push your arrow cursor against the top edge of the screen to make the menus slide back into view, or to the bottom edge to get the Dock back.

You use the same technique to get *out* of full-screen mode: Move the cursor to the top edge of the screen. When the window-resizing buttons

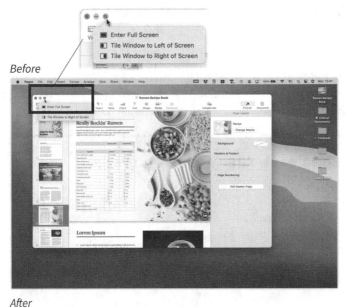

Before

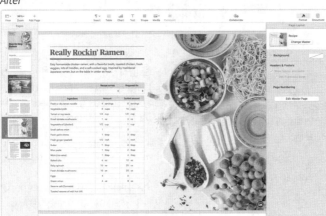

After

Full-screen mode

slide into sight, click the green dot again. This time, when your cursor tip approaches, it looks like this—⊖—which means "exit full-screen mode."

- **Almost full-screen (⊕).** The concept behind full-screen mode is solid: getting all the decorative window-control surfaces out of your way. But for many people, losing the menus, scroll bars, and Dock can feel disorienting and inefficient.

Fortunately, a compromise view is available: It makes the window expand to fill your entire screen *without* hiding the menus and the Dock.

To sample this view, click the green button *while you're holding down the Option (Alt) key* on the bottom row of your keyboard. (You'll know you're doing it right if a + symbol appears within the green button as you hover over it.)

In most non-Apple programs, Option-clicking like this instantly enlarges the window to its maximum size. In Apple apps, Option-clicking often produces a tiny menu; click the option called **Zoom**.

BRINGING BACK THE FOLDER PROXY ICON

For decades, advanced Mac fans enjoyed the presence of a tiny disk or folder icon in the title bar of every desktop window. It was called the *folder proxy icon*. (This is what happens when you let engineers name things.)

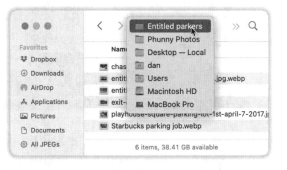

You could do two useful things with this icon. First, you could use it as a handle for the entire window, for dragging to another disk, to another folder, or even into the Trash. (Ordinarily, you can't drag an *open window* into these other places—only its disk or folder icon.)

Second, the folder proxy icon contained a secret menu, showing the hierarchy of folders that contain this window.

For example, if you're looking at a Finder window called **San Francisco**, this menu might reveal that it's inside a folder called **California**, which is inside a folder called **United States**. And you could jump to any of those outer folders by choosing its name in the secret menu.

In Big Sur, Apple made this feature even *more* hidden: The folder proxy icon no longer appears. But you can bring it back! Just click the window's name. Once the little icon has appeared, you can right-click or two-finger click it to view the secret folder-hierarchy menu. ✦

And, unfortunately, in a few particularly unenlightened apps, Option-clicking does nothing.

The Sidebar

At the left side of every Finder window is a column that's something like a table of contents for your entire Mac universe. These are the files, disks, and folders that either you or Apple have decided are important bookmarks.

Usually, you'll find four headings in the sidebar: **Favorites**, **iCloud**, **Locations**, and **Tags**. These are *collapsible* headings, meaning you can hide or show the list beneath each heading as suits your need for screen real estate at the moment.

Drag any icon directly onto the sidebar…

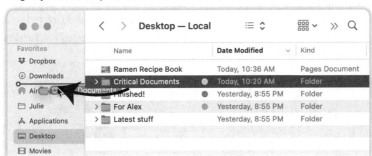

…to install it there.

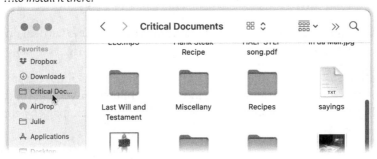

Adding a folder to the sidebar

To do that, point to one of the headings without clicking; click the tiny **Hide** or **Show** button that appears. Here's what you'll find beneath those headings:

- **Favorites.** Apple starts you off with the names of important folders on your Mac like **Downloads, Applications, Documents,** and **Pictures.** But you can add any icons you like—disks, files, programs, and folders you'll want to access often. Click anything in the Favorites list once to open it.

  You can rearrange these items by dragging them up or down. To remove an icon from Favorites, drag it a few inches away from the list, and then release your mouse button or trackpad. (You haven't actually deleted anything from your Mac; you've just removed its bookmark from the Favorites list.)

- **iCloud.** There's only one icon in this heading: iCloud Drive, which you can think of as a backup disk in the sky. Anything you put onto this drive is available from any of your other Macs, iPhones, or iPads, anywhere you can get onto the internet. See page 105.

- **Locations** lists the rest of the disks, cards, and other storage modules your Mac can access: hard drives, USB flash drives, CDs or DVDs, memory cards, other computers on your network (Chapter 14), and even iPhones and iPads you've connected to the Mac. (Your main hard drive doesn't start out listed here, but why not add it? Choose **Finder→Preferences,** click **Sidebar,** and turn on **Hard disks.**)

NOTE: As you point to the name of each removable disk (like a memory card, flash drive, or external hard disk), it sprouts a little ▲ logo, which means **Eject.** It doesn't actually make the disk pop out of the computer; instead, it just removes the disk's *icon* from the screen, which indicates that it's safe for you to *physically* remove the disk.

- **Tags.** This section lists all the color-coded keywords known as tags, which you can use to label your various icons. See page 116.

You can adjust the width of the sidebar by dragging the fine vertical divider at its right edge. (If there's a vertical scroll bar, drag the far-right edge of *that.*)

It's actually possible, by the way, to make the sidebar so skinny that it disappears entirely. If you're not a sidebar kind of person, you may like it that way. But if you change your mind, you can bring the sidebar back by choosing **View→Show Sidebar.**

The Finder Toolbar

There's a row of buttons at the top of every Finder window, too, which Apple ingeniously calls the toolbar. Most of them are dedicated to changing how the window's contents look—how they're displayed, grouped, and sorted. But the toolbar also harbors the search box, which can be useful indeed (page 120).

When you choose **View→Customize Toolbar**, you're invited to edit the toolbar—to add some buttons, remove some, or move them around. Some of the available options don't appear on the factory-installed toolbar and might be worth installing.

> **TIP:** If you make a real mess of things, you can always restore the toolbar to the way Apple originally had it. To do that, drag the horizontal bar (the "default set") out of this window and onto any open Finder window.

Drag buttons from the catalog of options directly onto the toolbar.

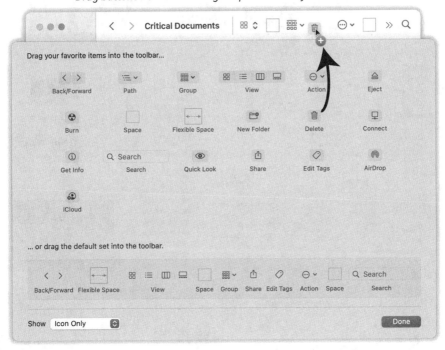

Editing the toolbar

Maybe the most useful element in the Customize Toolbar window, though, is the little pop-up menu at the bottom titled **Show**. It lets you choose how the toolbar icons look: little icons, as usual; word buttons (**New Folder**, **Delete**, and so on); or icons *with* words. The icons save the most space, but the words are least cryptic.

Four Views

When you open a window full of files, do you want to see them in a neat list? Or would it be more helpful to see them as miniatures of whatever pictures, text, or videos they represent?

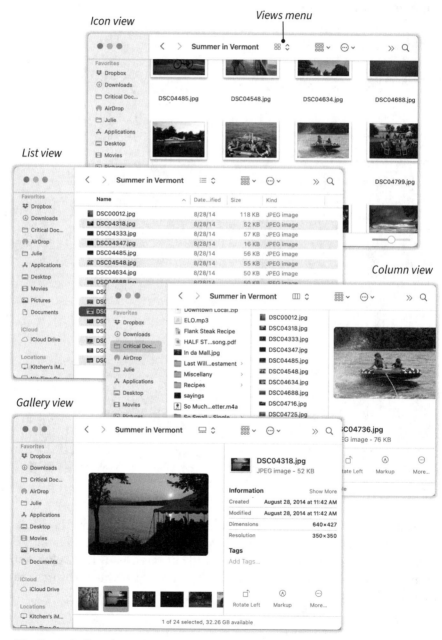

The four window views

The answer depends on what's in the window and what you're looking for. Fortunately, the Finder gives you four different views of any window, each suitable for a different kind of browsing. Every window remembers its own setting.

To switch among the views, click one of the four corresponding icons in the window's toolbar. Or, if you prefer, use the first four commands in the **View** menu: **as Icons**, **as List**, **as Columns**, or **as Gallery**. Here's a rundown of what these four views can do for you.

Icon View

In this view, every file, folder, and disk appears on the screen as an icon. You can make the icons bigger or smaller by dragging the size slider in the lower right of the window. (If you don't see it, choose **View→Show Status Bar**.)

Drag me.

Icon-size slider

Icon view is especially useful when you're looking over *documents*, because each icon is a miniature—a thumbnail image—of what you'd see if you actually opened those documents. A photo looks like a little photo. A video looks like a little video—and when you point to it even sprouts a Play button (▶) so you can watch it on the spot. A PDF file, or a Keynote, Pages, or PowerPoint presentation, offers little page buttons (◐ and ◑) that let you examine its pages right on the icon.

You can page through a PDF right on the icon.

Icon previews

For any window in icon view, you can choose **View→Show View Options** to open a dialog box full of further tweaks. You can, for example, change the icons' sizes; sort them alphabetically, by size or date, and so on; adjust the tightness of their grid spacing; make the icon text labels bigger or smaller; and specify whether each file's name appears beneath or to the right of its icon.

> **TIP:** Once you've made adjustments in this View Options panel, you can click **Use as Defaults** to apply these settings to all *other* windows you view in icon view.

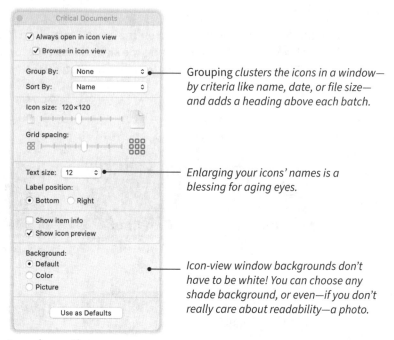

Grouping *clusters the icons in a window—by criteria like name, date, or file size—and adds a heading above each batch.*

Enlarging your icons' names is a blessing for aging eyes.

Icon-view window backgrounds don't have to be white! You can choose any shade background, or even—if you don't really care about readability—a photo.

Icon view options

If your icons look messy within the window, you can tidy them up:

- **Sort them temporarily.** Use the **View→Clean Up** command to make all the icons in this window snap neatly to an invisible grid. (Or use the **View→Clean Up By** submenu to sort them—by **Name, Size, Date Last Opened**, or whatever—in the process.)

  Even after the icons have been tidied, you're still free to drag them into a mess again.

- **Sort them permanently.** Use the **View→Sort By** command to make all the icons in this window snap to an invisible grid sorted by **Name, Size, Date**

Last Opened, or whatever—and *stay there.* You're no longer free to drag these icons into any positions you want; you've told the Mac to keep them sorted on a grid and, by golly, that's how they'll stay. (Until, of course, you choose **View→Sort By→None**.)

List View

In any window with a lot of stuff in it, list view is usually the logical choice. It shows everything in the window in a neat, sorted table, revealing not just each file's name but also its date, size, kind, and other details.

See those column headings, like **Name, Date Modified, Size**, and so on? They're actually buttons. Click one to sort the entire window by that criterion; click again to reverse the sorting direction.

The little ∧ or ∨ next to the column's name indicates which direction it's sorted. If you sort chronologically, for example, you might want the oldest files at the top, or the newest.

> **TIP:** This business of clicking a column heading to sort the entire window is a standard Mac technique. You'll also use it in Mail, Music, and anywhere else fine window contents are listed.

A flippy arrow (**>**) appears next to each folder in a list-view window. Click the arrow to expand the folder's contents as an indented sublist.

WHEN FOLDER SIZES APPEAR AS DASHES

One handy advantage of list view is that it's very easy to see the *sizes* of all the files in the window. But often, for *folders*, the Size column shows nothing but a couple of dashes.

That's easy to fix. Choose **View→Show View Options**. In the dialog box, turn on **Calculate all sizes**. Little by little, you'll see the actual folder-size readouts appear on the screen, as the Mac tallies the sizes of the files inside them.

In folders containing many thousands of items, it may take the Mac some time

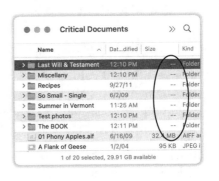

to compute these totals. But whatever it comes up with will certainly be more informative than "--." ★

Flippy arrow

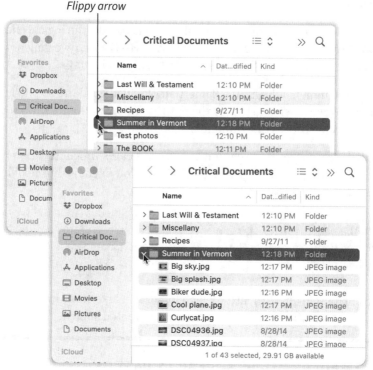

Using flippy arrows

You can have all kinds of fun organizing a list-view window. For example:

- **Choose different columns.** You can choose *which* columns of information you want to see by choosing **View→Show View Options** and turning on the appropriate checkboxes. (While you're at it, you can change the icon size, text label size, and sorting method for this window.)

- **Rearrange the columns.** You can use the column headings—**Date Modified**, **Size**, and so on—as handles. By dragging them horizontally, you can rearrange the columns. (The **Name** column is always leftmost.)

- **Change the widths.** You can make each column wider or narrower, too. Position your cursor carefully on the divider line between column headings, and drag right or left.

Column View

You probably won't use column view much, but every now and then it's a useful tool against window clutter.

The idea is to show you the contents of folders within folders as successive lists—columns—within a single window.

TABBED WINDOWS IN THE FINDER

In a web browser, you can have a bunch of web pages open simulta- neously—all in a single window—thanks to the invention of *tabs*. You click a tab to switch to a different web page you've opened, without having to fuss with opening and closing individual windows.

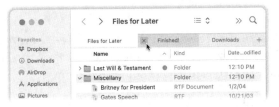

Apple has brought the same feature to desktop windows. Thanks to tabs, you can keep open the windows of several different disks or folders in a single window, making it easy to drag icons between them. Tabs can be a useful tool for juggling several windows at once; for everybody else, they represent a com- plexity that you can safely ignore.

To give desktop tabs a try, open a Finder window. Now choose **File→New Tab**, or press ⌘-T.

Just beneath the Finder toolbar, you can see you now have two filing-folder tabs. To add more tabs, click the **+** button at the far right (or repeat the **File→New Tab** business).

At this point, you can drag an icon from one tab to another. Or you can view the same window contents twice on different tabs—one in list view, one in icon view. Or view two different parts of a very long list in side-by-side tabs.

Feel free to drag the tabs horizontally to rearrange them. You can switch among tabs either by clicking on a tab or by pressing Control-Tab. Or drag a tab away from the others to turn it into a tradi- tional standalone window.

To close a tab, move your cursor to its left end and click the **✕** button that appears. (If you press the Option key at the same time, you close all tabs at once.) ★

The first disk or folder you open is always in the far-left window, displayed as a list. If you click a folder in that list, its contents appear in a new skinny column to the right. If you click a folder in *that* list, you see its contents in a third column. And so on.

This lets you navigate a chain of folders without losing your place, making it far easier to backtrack if you went down the wrong rabbit hole.

The wider you make the window, the more columns you can see without scrolling. You can also adjust the widths of the individual columns by dragging the vertical dividing line between them.

> **TIP:** If you right-click or two-finger click the column divider, you get a secret shortcut menu containing useful commands for adjusting the column widths. You can choose **Right Size This Column**, for example, which makes the column exactly wide enough to fit the name of the longest item in its list. **Right Size All Columns Individually** does the same for *all* the columns in the window. And **Right Size All Columns Equally** makes all the columns the same width—based on the width necessary to contain the longest filename in any one of them.

As usual, you're welcome to choose View→Show View Options for a column-view window. In the resulting dialog box, you can adjust the text size, hide or show the icons in each column, and so on.

Gallery View

In this view (see "The four window views" on page 96), you get a gigantic, full-window preview of whatever icon you've highlighted in the scrolling row at the bottom.

That's a super-useful arrangement when you're perusing a folder full of photos, movies, presentation files, or documents, because you get to see what's in each one at a readable size. (Gallery view is not, however, very useful in most other situations, because it dedicates so much space to the icon preview.)

Previews and Quick Looks

One of the coolest features on the Mac—something Windows has never been able to match—is the ability to show you what's in a file without your actually having to open it. You can view a photo, play a movie or sound, read a document (Pages, TextEdit, Microsoft Word, PDF), page through a slideshow

(Keynote or PowerPoint), check out the pages of a spreadsheet (Excel or Numbers), and so on, right in the desktop window.

The Finder offers two ways to avail yourself of these sneak peeks.

Preview Pane

In any of the four window views, you can make the preview pane appear by choosing **View→Show Preview**. That's a pane at the right side of the window that displays an instant preview of any icon you click in the window. It also reveals some details about the file; depending on the file type, you might see its creation date, file size, number of pages, or photographic details.

Page through a PDF, or play a music or video file.

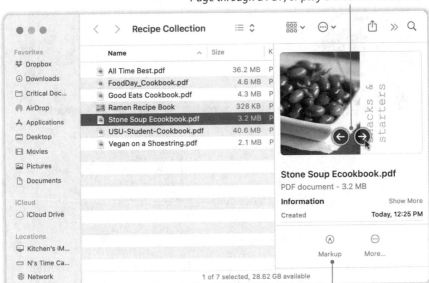

The preview pane

Annotate a PDF or graphic.

Quick Look

Quick Look is fantastic. It's a window that pops open when you tap your space bar, showing the preview of what's in a file you've highlighted.

Yes, tap your space bar. That's it. No menu, no clicking—just select an icon in the Finder and then tap space. Instantly, you're browsing a document at nearly full size, without having to open any app at all. (You close the Quick Look window with another press of your space bar.)

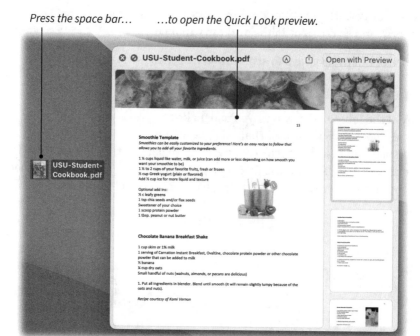

Press the space bar...   ...to open the Quick Look preview.

Quick Look

TIP: Ordinarily, you're not allowed to open a file you've put into the Trash (page 119). But you *can* use Quick Look on files in the Trash—for example, to make sure it's not something important that you're about to nuke forever.

Quick Look is filled with quick tips:

- **Once you've opened the Quick Look window,** you can click another icon, and another, and another (or press your arrow keys to walk through the files in the window). The Quick Look window changes to show you what's in each one.

- **Every Quick Look window has** a ⊘ button at the top left. Click it to enlarge the document preview for greater legibility.

  To exit full-screen mode, click the ⊕ button.

- **If you've selected a bunch of icons** and *then* entered full-screen mode, you can click ▸ to begin a full-screen slideshow. It's great for looking over a bunch of photos you've just downloaded.

iCloud Drive

What Apple calls iCloud is a big suite of mostly free internet-based services for Apple customers. You can read all about them in Chapter 7.

One of the most important iCloud features, though—and one that's built right into the Finder—is the iCloud Drive. There's an icon for it right there in your sidebar, looking like a little cloud (⌒).

You can think of it as a magic folder: Anything you drop into it appears instantly in the iCloud Drive folders of all your *other* Apple stuff—Macs, iPhones, iPads. (Actually, you can even access your iCloud Drive from a Windows PC.)

> **NOTE:** Have you ever used Dropbox? This is Apple's version of the same thing.

All the iCloud Drive's contents are also accessible on a website—iCloud.com, of course. That's handy, because it means you can grab your files even if, in a pinch, you're forced to use somebody else's computer (::shudder::).

Thanks to iCloud Drive, you never again have to email files to yourself, copy them to a flash drive to take home, or worry that you'll lose important files forever when a lava sinkhole opens up and swallows your Mac. Those files are still safe "in the cloud" (that is, in an internet-connected data center somewhere).

This icon means "I'm not really on your Mac, but I'll download on demand."

Inspecting your iCloud Drive

THE MAC HELP MENU HELPS YOU SEARCH THE MENUS

It's been many years since Apple included a printed user guide with its products. That's probably one reason you picked up this book.

Apple hasn't ignored the issue of instructions completely, though. There is, of course, a Help menu in the Finder.

If you select macOS Help from that menu, you open a little Help program that works something like a web browser. It's awfully terse, basic, and humor-free, but hey—you can never misplace it.

Actually, one of the best features of the Help menu is the search box at the top. When you type something into this box, the list of search results below it doesn't just include help screens. It also lists any of the current app's *menus* that contain the word you searched for.

Suppose, for example, that you're using Photoshop, a program with over 500 menu commands. You recall that somewhere in there is the **Blur** command. But are you really supposed to hunt through

12 menus (and dozens of submenus) to find it? It'd take you all year.

There's no need. Just open the Help menu and, in the search box, type *blur*. The search results include any menu commands that contain that word. Best of all, if you point to one of these without clicking, the actual Photoshop menu containing that command opens for you, and a bright-blue arrow points to the command.

This concept is especially useful in a web browser like Safari. You know the History menu, which lists the last several hundred websites you've visited? You can use the Help menu to search everything inside it.

Suppose you remember visiting a page about the duck-billed platypus—but you don't really feel like trawling through 7,000 items in your History menu to find it. No problem: Use the Help-menu trick!

It's the greatest feature nobody's ever heard of. ✦

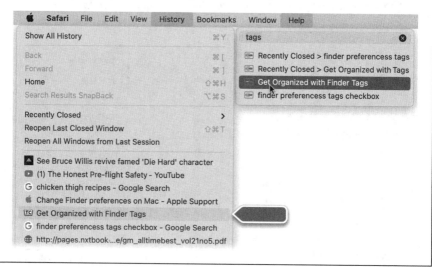

iCloud Drive also makes a fantastic backup disk—or it would if it were bigger. Apple gives you only 5 GB of storage space for free. If you like, you can pay a monthly fee for more space: 50 gigabytes ($1 a month), 200 gigabytes ($3), or 2 terabytes ($10), which is 2,000 gigs.

Sharing Files and Folders

You're not the only one who can access what's on your iCloud Drive. In times of collaboration, you can also share certain files and folders on it with anybody across the internet. Well, at least anybody with an iCloud account (also called an Apple ID), which is free.

This iCloud Drive sharing business is incredibly useful. It means you and your friends/family/co-workers can work on documents, or folders full of them, wherever they happen to be and at whatever crazy times of day or night they feel inspired.

To get started, open your iCloud Drive. Click the file or folder you'd like to share. From the 🖅 icon on the window toolbar, choose **Share File** or **Share Folder**.

Sharing an iCloud Drive folder

Now you have three questions to answer:

- **How do you want to invite your collaborators?** The invitation to access your file or folder looks something like a web link, which the Mac is now proposing to send to your invitees. It gives you four options: **Mail** (send the link in an email), **Messages** (send the link as a text), **Copy Link** (ready for pasting into any app you like), or **AirDrop** (send to a nearby Mac, iPhone, or iPad wirelessly, as described on page 333).

- **Who's allowed to use the link?** If you choose **Anyone with the link**, then anybody who knows the link can look at what you're sharing. That might be just what you want if it's an ad for your next band performance or an essay about a cause you believe in. But if you choose **Only people you invite**, then access is limited to the precise set of people you're about to email or message.

- **Are they allowed to edit, or just read?** The **Permission** pop-up menu offers two choices. If you choose **Can make changes**, then the people you've invited can edit your shared document, or add and remove files from your shared folder. All collaborators see the same changes. If you choose **View only**, then it's "look, don't touch."

Once you set up your responses, click **Share**. You're now asked to enter the email or text address of your recipients, or choose them via AirDrop; when you click **Send**, the deed is done.

Once they click the link you sent, they'll be able to edit (or just read) your shared documents; for a shared folder, they're free to see what's inside, or (if you permitted editing) they can add and remove files and make changes to the documents inside.

At any point, you can revoke permission for somebody, change what kind of permission they have, or shut down the whole thing. To do that, click the shared file or folder, click the 🖐 icon on the toolbar, and choose **Manage Shared Folder**.

You get a little window that contains the same **Who can access** and **Permission** pop-up menus, a **Stop Sharing** button, and a list of all the invitees. Click the ⋯ button next to a name to change their access or remove it altogether.

CHAPTER FIVE

Finding and Organizing Your Stuff

All those folders and windows described in the previous chapter are really only stages for the featured players: the thousands of *icons* that represent the apps you use and the documents they produce.

Fortunately, it would be hard to imagine an operating system offering any more ways to manage, manipulate, move, copy, rename, search, and delete them.

How the Mac's Folders Are Set Up

A Mac is designed to be used by different people at different times. You and your family members, or co-workers, or fellow students, each have your own password. You each sign in to find your own files, bookmarks, and settings.

That arrangement may not be *your* situation—if you own a MacBook laptop, for example, you may well be the only person who ever uses it. But the multiple-user option does explain some of the Mac's most mystifying aspects.

For example, consider how its folders are organized. Treat yourself to a little tour, starting by choosing **Go→Computer**. The resulting window displays an icon for your Mac's built-in hard drive.

> **NOTE:** On most Macs, the "hard drive" isn't actually a moving spindle of rotating platters, as in the days of yore. Instead, it's just a memory storage chip (an *SSD*, or solid-state drive, to be exact). But everybody still thinks of the storage unit inside the Mac as a hard drive. Including Apple—that's why yours came from the factory bearing the name **Macintosh HD**.

If you double-click **Macintosh HD**, you reveal at least four folders: **Applications** contains all your apps (programs). **Library** and **System** contain the operating system itself; most people never have any need to open these folders.

And then there's the **Users** folder. This is the one you care about.

Inside is one folder for everybody who uses this Mac. It could be that there's only one folder here, named for you. Maybe there's one for each family member. Maybe there's one for every kid in the school.

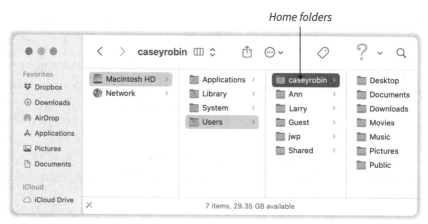

The Mac folder structure

In any case, these are the *home folders*. Your home folder, which bears your name, stores your entire Mac universe: all your files, your work, your music and movie collections, your web bookmarks, your address book, your email, and all the settings that make the Mac behave the way you like it.

Whenever you're lost on the Mac, you can always return to your own home folder by choosing **Go→Home**.

> **TIP:** It's useful to put your home folder into the sidebar, so its icon is available in every Finder window. To do that, choose **Finder→Preferences**, click **Sidebar**, and turn on the ⌂ checkbox that bears your name.

Each home folder comes with a standard set of folders that help you organize your digital life:

- **Pictures, Music, Movies.** These folders contain exactly what they say. The Photos app (page 282), for example, keeps all your photos in the Pictures folder.

- **Documents.** Suppose you're using an app that produces documents: text files, spreadsheets, presentations, or whatever. The first time you use the Save command, the app will probably propose storing your new document in this folder, making it easy to find it again later.

- **Downloads.** Any file you download from the web, any attachment you save from an email, or any file you receive by AirDrop (page 333) lands here, for easy retrieval.

- **Desktop.** Some people like to leave their most important files lying loose on the desktop, not in any folder. Behind the scenes, though, those files actually sit in this Desktop folder. The icons you see are just a visual fiction the Mac creates for your convenience.

- **Public.** You can safely ignore this folder until the day you add your Mac to a network and wish to share files with other people on it. Chapter 14 has the details.

- **Applications.** The apps on your Mac usually sit in the Applications folder. But, believe it or not, *this* Applications folder is not *that* Applications folder. This one is designed to hold apps that only *you* get to use, not everybody on your Mac. It's probably empty.

You can create your own folders in your home folder, too—and you should. Over time, your apps may create folders of their own here, too. But the point is that this one folder contains every shred of software related to *your* activity on the Mac.

> **TIP:** In the Finder, you can jump directly to the most important folders with a key combo. You can get to the Home, Applications, Desktop, or iCloud Drive folder by pressing Shift-⌘-H, A, D, or I, respectively. All you have to memorize is their first initials. The rest of the keystroke is always the same: Shift-⌘.
>
> You can get to the Documents folder with Shift-⌘-O, too. Now, O is not the first letter of "Documents"—it's the second—but the "D" was taken by Desktop.

Naming and Renaming Things

Everything on your Mac—every file, folder, disk, app—has a name.

To rename something, click once on its name or icon and then press Return. (Or right-click or two-finger click it, and then choose **Rename** from the shortcut menu.)

A renaming rectangle appears, and the existing name is highlighted. At this point, you can rename the icon just by typing. You're allowed to use up to 255 characters—any letters, numbers, or symbols you want except the colon (:). Also, no two files can have the same name *in the same folder*.

When you're finished typing, press Return again, or just click somewhere else.

Selecting Icons

At the very heart of the graphic interface that the Mac made famous, there's a noun-verb structure. You select something (like an icon), and then act on it (like **Move to Trash**).

Mastering this technique, therefore, requires that you know *how* to do the selecting. Obviously, to select a single icon in the Finder, you click it once. But how would you select more than one item at a time?

Like this:

- **Select everything in the window.** Choose **Edit→Select All**.

- **Select neighboring icons.** Drag diagonally (start your dragging in a blank spot) to highlight a few nearby icons.

Drag diagonally.

Drag to select icons

- **Select consecutive items in a list.** In list or column view, you can drag vertically over the files' names to select them. Or click the first item, and then Shift-click the last one. Everything in between gets selected.

RENAMING BATCHES OF FILES AT ONCE

Suppose you're writing a novel that you're confident will be a best-seller when it's finished.

At the moment, you've got 37 chapter files in a folder. Their names are *The Hobbit Chapter 1*, *The Hobbit Chapter 2*, and so on.

And then, to your tremendous annoyance, the publisher claims that somebody else once wrote a book with the same title. They suggest that maybe you should change yours.

It would take you forever to rename those files one at a time!

Fortunately, the Mac offers a *batch renaming* feature, which lets you rename entire clumps of icons in one fell swoop.

Begin by selecting all the files, using the techniques described on the facing page. Now right-click or two-finger click any one of the selected files; from the shortcut menu, choose **Rename**.

The box that appears offers a pop-up menu containing three powerful options.

If you choose **Replace Text**, you'll search-and-replace text within all the files names. In one step, you could rename *The Hobbit chapter 1*, *The Hobbit chapter 2*, and the other files to *The Small Barefoot Explorer chapter 1*, *The Small Barefoot Explorer chapter 2*, and so on.

If you choose **Add Text**, you can append some new text onto every filename—at either the beginning or the end of it. You could add "2021" to every file in your Dreams Still Unattained folder, for example.

The third option, **Format**, completely replaces the files' existing names with a new, consistent naming scheme. There's a base name—whatever you type into the **Custom Format** box—plus a number or a time and date.

Suppose, for example, that you type *Worries List* into the **Custom Format** box; that's your base name.

If you choose **Name and Index** from the second pop-up menu, the resulting filenames will be differentiated by a number, which appears either before or after the base name. Example: *Worries List 1*, *Worries List 2…*

Name and Counter is exactly the same, except that the Mac pads out the number to five digits. Example: *Worries List 00001*, *Worries List 00002…*

Finally, **Name and Date** tacks the current time and date to the end of every file's name. Example: *Worries List 2021-5-19 at 11.14.04 PM.*

Of course, this means all the files would have exactly the same name—and, sure enough, you'll get an error message if you use this option unless the original files are all in different folders or have different file types (.jpg, .pdf, and so on). ★

Click the first icon...

⌘-click to remove one from the selection.

Shift-click the last one.

Selecting consecutive files

Once you've highlighted a bunch of icons, you can apply your *verb* to all of them simultaneously. You might drag them as a group to another folder or another disk. You could copy them all. Or you could choose **File→Move to Trash**.

> **TIP:** Your keyboard offers a great way to find a file in a haystack. Suppose you're looking for a document called Recipes.pdf in a folder teeming with hundreds of documents. Just type *the first letter or two* of the file you want—*re*—to make the Mac jump right to it, highlighting it.

Moving and Copying Icons

Now that you know how to select an icon—or several—you can put that knowledge to work. You can move or copy them from their current window to another using one of two methods:

- **Drag them.** If you drag the highlighted icons onto a different folder on your hard drive, the Mac moves them. If you drag them onto a different *disk*, however, the Mac *copies* them.

> **TIP:** And what if you want to *copy* them on the same disk? It could happen. In that case, hold down the Option key just before you finish dragging. That tells the Mac, "Copy instead of move."

- **Copy and paste them.** The problem with dragging icons is that before you even begin dragging, you must arrange your windows to have the destination visible on the screen. Sometimes, that can require a bit of fiddling. The **Copy** and **Paste** commands, on the other hand, can do the job even when the origin and destination points are far away from each other.

Once you've selected the icon or icons you want to move, choose **Edit→Copy**. (Or press the universal keyboard shortcut for Copy, ⌘-C.) You've just placed copies of those icons on the invisible Mac Clipboard.

For step two, find and open the window where you'd like to put those icons. At this point, choose **Edit→Paste**. (Or press ⌘-V.) Like magic, the Mac places a copy of the original icons in this new location.

> **TIP:** If you hold down the Option key as you paste, you tell the Mac to *move* the icons instead of copying them. That is, they disappear from their original location.

Aliases

Now that you're familiar with the notion of copying your files, you're ready to handle the concept of *aliases*.

On your screen, an alias looks for all the world like a duplicate of the original icon. But the alias is only a *pointer* to the original, occupying virtually no disk space. When you double-click the alias, the original opens, wherever it is on your Mac.

The beauty of this system is that a single icon can seem to exist in multiple places simultaneously. (There is only ever one real file; the aliases are its shadows.)

> **NOTE:** At this point, you'd be forgiven for wondering why Apple has created so many different mechanisms for allowing you easy access to the icons you consider important. You know: Drag them to the Dock! Drag them into the sidebar! Leave them on your desktop! Do we really need yet another way to give favored icons special treatment?
>
> But remember: MacOS has evolved over the years. Aliases, introduced in 1991, were the very first solution to the "quick access to a file" problem. The Dock came along in 2000; the sidebar in 2003. When your mandate is to release a new version of your operating system every single year, adding new features each time, some redundancy is inevitable.

To create an alias of a highlighted icon, choose **File→Make Alias**. You'll see what appears to be a second copy of the original icon, with only a little arrow "badge" in the corner to identify it as an alias.

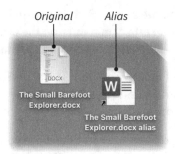

Original   Alias

Creating an alias

You can create as many aliases of an icon as you like, and you can put them in as many folders as you like.

> **NOTE:** So what happens if you trash the original? Or maybe the original is on some flash drive in an office drawer, and you're now at home? In those cases, when you try to open the alias, you'll just get an error message that the original can't be found.

Tagging Your Files

Over the years, it became clear to hard-core Mac fans that putting *names* onto their icons wasn't always enough. What if you just want to round up all your daughter's schoolwork, no matter what those files were called? What if you want to find the *first drafts* of everything you've written? What if you want to collect all the files on your entire computer that have to do with the Higginbotham Proposal, no matter what folders they're currently in?

That's the idea behind *tags*. A tag is a color-coded category name that you can slap onto any icon, like *daughter*, *first draft*, or *Higginbotham*. You can apply multiple tags to a single icon, too, in the unlikely event that your daughter created the first draft of the Higginbotham proposal.

To set up the tags you want to use, choose **Finder→Preferences**; click **Tags**.

Create Your Tags

As you can see, Apple starts you out with a set of tags, named only for their colors. But you can rename one (click the existing name twice slowly), change its color (click the existing color dot), create a new tag (click **+**), or delete a tag (select it and then hit **−**). If you turn on the blue checkbox next to a tag's name, it will appear in the list of tags in the Finder sidebar for quick access.

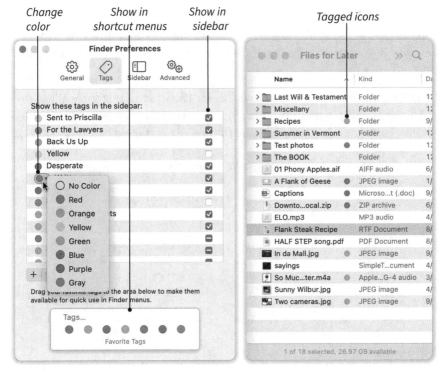

How to use tags

Apply Tags

There are about as many ways to apply tags to your files as there are songs on Spotify, but here are some of the most useful:

- **Use the File menu.** Select the icons. Choose a tag color from the **File** menu.

> **TIP:** If you choose **File→Tags** instead, you'll have the opportunity to apply several tags at once.

- **Use the Finder toolbar.** Select the icons. Choose a tag from the ⬦ icon on the Finder toolbar.

- **Use the sidebar.** Drag the selected Finder icons *onto* the name of a tag at the bottom of the sidebar.

> **TIP:** You can also apply a tag to a file at the moment of its creation—in the Save dialog box described on page 140.

Use Your Tags

Suppose you've patiently trawled your Mac, painstakingly applying appropriate tags to every icon you might ever want to see again. Here's where it all pays off:

- **Click a tag color in the sidebar.** Boom: The window instantly fills with all the icons on your entire Mac that bear that tag, no matter what folders they're actually in. One click, and they're all assembled before you.

Click a tag to round up those icons from everywhere.

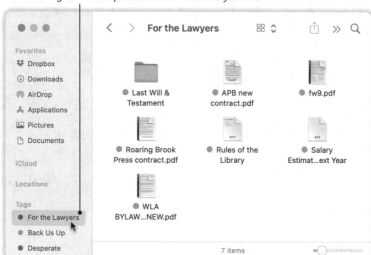

Sidebar tag roundup

- **Use the search box in every window.** You know the search bar at the top right of every Finder window? It can search by tag, too.

- **Use Spotlight.** You can also limit a search to files of a certain tag when using Spotlight, the Mac's global search feature (page 120). Just begin your search by typing *tag:first drafts* (or whatever the tag's name is). You can then go on to refine the search by typing *additional* search criteria, as described on page 128.

Removing Tags

You can *detag* an icon, too. Highlight it, choose **File→Tags** (or open the ◇ menu), and delete the tag name from the little window that appears.

The Trash

If you really inspect what Apple calls the Trash (the icon at the right end of the Dock), you'll realize that *it's not actually a trash can*. It's a wastebasket.

Take whatever time you need to get over the great lie.

The Trash

Whatever it's called, you use it to delete files and folders from your Mac. Just drag the condemned icons until your arrow cursor is right on top of the Trash icon, and then let go. (The wastebasket on the Dock cutely overflows to show that there's something inside.)

THE INFO IN THE GET INFO BOX

In any standard list view, you can see useful information about every file in the window: name, date, size, tags, and kind. That may seem like a lot of tweaky detail, but believe it or not, it's only the tip of the tip of the iceberg. To see the *full* dossier of details the Mac knows about a file, choose **File→Get Info**.

The tall, skinny box that appears is packed with details about the icon you selected. What app opens when you double-click it. The last time anybody changed it. Where it sits on your hard drive. What comments you've typed in for your own reference. Who's allowed to edit this document when you share it on a network.

If you've ever used Windows, you may be familiar with the Properties box for a file. Same thing.

You probably won't use Get Info often. But now and then, for troubleshooting purposes, somebody may ask you to "get info" on a file.

And now you'll know what they're talking about. ★

> **TIP:** If you have a gigantic monitor, it may be easier to choose **File→Move to Trash** instead of trying to drag the icons across the great barren tundra of your screen. Or just press ⌘-Delete.

Now, Apple realized early on that throwing something away and then *changing your mind* is a standard human foible. That's why, from the day the first Mac came out in 1984, it has offered a safety net.

You know how, when you throw something into the trash can in your kitchen, it's not gone *forever* until you empty the trash? You can still root through the garbage to rescue something.

It's the same way on the Mac. At any time, you can click the Trash icon on your Dock to open a window containing everything you've discarded so far. If you've change your mind about getting rid of something, select it and then choose **File→Put Back**. (Or press ⌘-Delete again.) The grateful icon magically leaps back into whatever folder it came from.

To take out the trash permanently, choose **Finder→Empty Trash**. If one final warning appears, click **Empty Trash** to complete the job.

> **TIP:** If you can do without that final confirmation box, choose **Finder→Preferences→Advanced** and turn off **Show warning before emptying the Trash**.

Searching for Stuff

Spotlight is the Mac's search feature. It can find any file on your computer—not just by name, but by its characteristics (you can search for "spreadsheets" or "photos") or even the *words inside* a document. You can even type in vague searches like *photos from January* or *files I worked on last year*, and Spotlight will deliver what you ask for.

Over the years, Spotlight has grown in power and ambition. Today, it can also search the internet. It can search the dictionary. It can show you what movies are playing in theaters. It can show you, on a map, the location of any address. It can fetch the latest sports scores, stock quotes, and weather reports. It can do math for you. It can make unit conversions, like meters to feet, Celsius to Fahrenheit, or dollars to euros.

The hardest part of using Spotlight, in fact, is just *remembering* to use it the next time you need information from any of those sources.

Starting a Search

The mousy way to begin a search is to click the Q icon on your menu bar.

The faster way is to press the all-powerful keyboard shortcut ⌘-space bar.

In either case, the Spotlight window opens. At the outset, it's more than a search box; it's basically an ad for all the different kinds of things Spotlight can find, to help you remember its seething power.

At this point, begin typing whatever you're looking for. With each letter you type, the results list refines itself.

> **TIP:** You're welcome to move the Spotlight box, or even its results box, around the screen; use any blank spot as the top of the handle.

Big Sur's Radical Research Revamp

For many years, Spotlight's list of search results came grouped by category: **Apps, Documents, Folders,** and so on. And there was a preview pane: If you clicked one of the search results, it displayed a miniature of the document or web page, a movie trailer, a map of an address you sought, the name/address card of somebody whose name you typed, a complete dictionary definition, a five-day weather forecast, or whatever.

In Big Sur, things aren't so simple.

At the top of the results, you usually get the result Apple thinks you're *most likely* to be seeking, no matter what its category. If it's at all obvious what you're looking for, it appears here.

For example, if you type *keyn,* the Mac assumes you want the Keynote app. If you type *Hrnezy,* and that's the name of the main character in the novel you're writing, your manuscript will be at the top of the list. For stock symbols (*AAPL*), weather (*Miami weather*), and sports (*Cavaliers*), this top line even includes an up-to-date readout of the stock price, weather report, or latest score.

If one of these results is indeed what you're looking for, click it (or, if it's highlighted, press Return) to open it.

> **NOTE:** In general, clicking a top hit opens it immediately. But if there's a **>** off to the right, then a preview is available. One click opens the preview; a double-click opens the website or whatever it is.

The Spotlight search box

Preview panel is available.

First section: top hits (if any)

Second section: search variations, on the Mac or the web

Other results, by category: Click once to see a preview, twice to open.

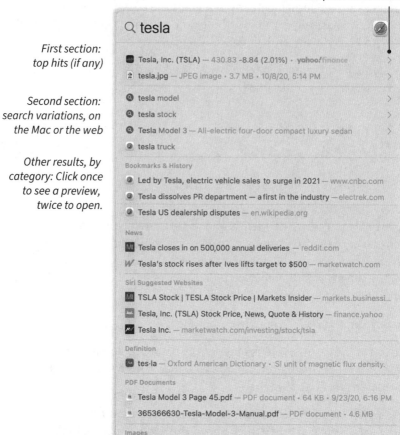

How Spotlight works in Big Sur

The next chunk of results, also unlabeled, consists of auto-suggested, click-able variations on your search term. If you've typed *chicken*, these sugges-tions might include **chicken recipes**, **chicken satay**, and so on.

Each one displays either a tiny ⌕ icon or a Safari compass icon. This is Spotlight trying to understand: Are you searching your *Mac* or searching the *web*? If you click one of the ⌕ items, you perform a new search, using that entire phrase as the search term (*chicken recipes*, for example). If you click one of the Safari items, Safari opens and performs a search for that term.

The remaining search results work as they did in previous macOS versions: They're grouped by category, like **Bookmarks & History, News, Definitions, PDF Documents, Documents, Mail & Messages**, and so on.

Here's the important part: *Click once* on one of these results to open the traditional preview pane that illustrates the currently highlighted result (see "Searching Notes" below). Double-click to open the website, document, or whatever it is.

Mac Searches

Honestly, what you'll probably use Spotlight for most often is searching your Mac. It's incredibly smart; you don't even have to remember the *name* of the file you're looking for. Here are some of your options:

- **Search by name.** You can type only the first few letters of the file you're looking for—type *fantas* to find your Fantasy Curling League.doc file, for example.

> **TIP:** This is a really great way to open an app in a hurry. You can press ⌘-space bar, type *safa*, and boom: You've highlighted Safari, which you can then open by pressing the Return key.

Preview pane

Searching Notes

- **Search inside the files.** Spotlight can even locate the *typed words inside files.* If you're certain that dehydrated kombucha was an ingredient in that amazing pasta recipe you made last year, but you can't remember the name of the recipe itself, use Spotlight to search for *kombucha.*

  Spotlight is perfectly capable of seeing into all kinds of text documents (Microsoft Word, TextEdit, email, PDF files), but it can even locate photos according to whatever text snippets are part of them, like their names or where they were taken.

- **Search by kind.** You can type *spreadsheet, PDF, presentation, JPEG, .doc, photos, music, disk image,* and so on.

- **Search within your Apple apps.** Spotlight can even see into your Notes, Reminders, Contacts, Mail, Calendar, Messages, and other apps that came with your Mac. This is a great way to look up some recipe in Notes, the details of an appointment in Calendar, or somebody's address in Contacts—without even opening those apps.

- **Search your memory.** Spotlight understands plain-English descriptions of the files you're looking for. You can type things like *email from last week, photos from August 2020, slides from last year containing WidgetCorp, files I created yesterday, emails from Eric Whitaker that contain documents,* and so on. You can mix or match any combination of file types, people's names, periods of time, and the contents of the files themselves.

> **TIP:** You can turn off certain categories of files, so Spotlight won't bother finding them in your searches. Open **System Preferences→Spotlight**, and turn off the checkboxes for categories you don't need—**Fonts**, **Movies**, or whatever.
>
> While you're here, you can also declare certain disks or folders off-limits from Spotlight searches. Click **Privacy**, and drag those private folders or disks directly into the list window. From now on, they're invisible to Spotlight.
>
> And who would ever *not* want Spotlight to see inside certain folders?
>
> You know who you are.

Web Searches

Of course you know how to search the web: Open your web browser, go to Google, type what you're looking for.

But that's twice as much work as using Spotlight. Not only can you use Spotlight without having to open a web browser, but you get your results right there in the search window.

ADVANCED DOCUMENT SEARCHING

Spotlight may seem simple enough—type what you're looking for, view the results—but it's hiding a lot of cool tips, especially when it comes to searching for files. All these tricks apply to the results in either top hits or the detailed results screen that shows results by category.

You can find out where a file is on your hard drive (what folder it's in). To do that, hold down the ⌘ key. You see the folder path at the bottom of the preview pane or (if the preview isn't open) appended to the file's name itself.

You can jump directly to a document, in whatever folder contains it, by ⌘-*double*-clicking its name in the results.

The files in the results are living icons. You can drag them right out of the results list—into a folder, onto the desktop, or into the Trash, for example.

You can use quotes just the way you do in Google. For example, if you search for *jumbo shrimp*, you'll find documents that contain both those words, but not

necessarily together. But if you add quotes (*"jumbo shrimp"*), you'll find documents containing that exact phrase.

You can limit a search to certain kinds of documents by typing *kind:* and then the type of file you want: *image*, *app*, *folder*, *email*, *movie*, *music*, *pdf*, *bookmark*, *font*, *spreadsheet*, *presentation*, or whatever. For example, to round up all pictures, type *kind:image*. You can then continue typing your search request—to specify the file's name or what's inside. For example, *kind:image Grand Canyon*.

You can also limit the search by time, using phrases like *date:yesterday* or *date:last week* (or *month*, or *year*).

And don't forget about tags (page 116)! Here's your chance to confine the search to files with, for example, *tag:family* or *tag:orange*. Here again, you can then continue typing to further refine the search among the items with those tags. For example, you could type *tag:family date:this month grand canyon* as one search. ✦

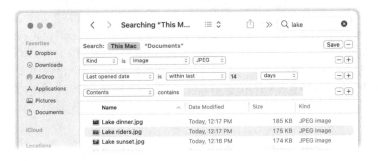

Anything you might type into Google, you can type into Spotlight. Information, breaking news, restaurants, businesses, movies, stock prices, weather, tweets, YouTube videos, sports scores, flight information—it's all right there.

And don't forget that when you click anything in the Search Suggestions list, you open the traditional Spotlight preview pane, where you can *see* the video, headline, movie information, stock graph, or whatever.

Calculations

You can type calculations into the Spotlight search bar. Use * to mean "times" and / to mean "divided by." For example, if you type or paste in *27*456.2-77+256*, Spotlight shows you the answer as the top hit: 12,496.4.

Spotlight even understands square roots—you can type *sqrt(81)* to get the answer 9—as well as functions like log(x), exp(x), sin(x), sinh(x), and e. It even knows what you mean by *pi*.

Spotlight can also convert units of time, temperature, length, mass (weight), volume, area, force, power, and international currency. Into the search box, just type what you want to convert, like *7500 feet*. The top hit guesses at the conversion you intended (**1.42 miles**); if that's not right, click it to see a preview pane filled with other conversions, like kilometers, yards, and inches (**90,000**).

The currency converter is pretty slick, too. Here again, you can just type the amount you want converted (*$1400*) to see Spotlight's best guess at the equivalent you want (in this case, it picks *euros*)—or you can specify the converted units, like *$1400 in yen*. Either way, Spotlight goes all the way to the internet to find the latest conversion rates.

Window Searches

Most people, most of the time, search the Mac using Spotlight. But there's also a search box in every single Finder window. (If the window is too narrow to show the entire box, you just get a Q button on the toolbar. Click to see the box.)

TIP: Or just press ⌘-F, which always makes the search box appear. (F for *find*, of course.)

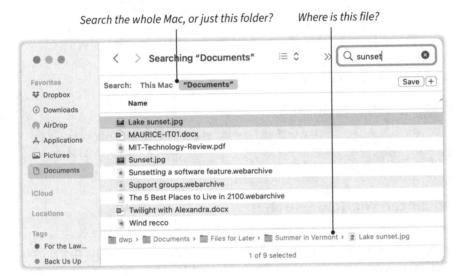

Search the whole Mac, or just this folder?   *Where is this file?*

The Finder search box

Now, if Spotlight is so great, you might wonder why the Mac needs another, duplicate search mechanism. Honestly, it doesn't—but the window-search method does offer certain distinctions:

- **It doesn't search the internet.** It searches just your Mac.

- **It can search only one window.** Spotlight always searches your entire Mac—but you can limit a Finder-window search to what's in the current window.

- **It can memorize a search.** Suppose that at the end of every week, you round up all the Word files with the tag *manuscript* that you've edited in the past seven days—and you copy them to a flash drive as a backup. Using the Finder window search feature, you can recreate this complicated search for reuse later with a single click.

The basic search goes just as you would expect: Type what you're looking for into the search box. As you type, the window shows you a list of matching files. (These are files whose names match what you've typed, or that *contain* words that match what you've typed.)

The list of search results acts like a standard Finder window. You can double-click a file in the list to open it, press the space bar for a Quick Look at it, drag it to the desktop to move it there, change the view (to icon view or column view, for example), sort the list (by clicking the column headings), or just close the window and forget the whole thing.

There's really only one more thing to learn about the basic search: It comes set to search the entire Mac, just like Spotlight. But the results window always includes at least two buttons at the top, called **This Mac** and whatever the window's name is. If you click the window's name here, you redo the search but limit its scope to the window you have open.

> **TIP:** You can make it so the Finder window search *always* searches the window you've opened. To set that up, choose **Finder→Preferences→Advanced**. Open the pop-up menu called **When performing a search**, and choose **Search the Current Folder**. Finder window searches are much more useful that way.

Advanced Window Searches

You can actually set up the Finder window search to be every bit as elaborate as Spotlight searches: You could find a file with the word *kumquat* in the title, created in June last year, with a size over 5 MB, that's a PDF document. Not one person in a hundred actually uses this feature, but because you may see pieces of it on the screen and wonder about it, here's how it works.

When you first type into the search box, the Mac tries to figure out what aspect of your files you're trying to search—and it offers you those choices in a little suggestions menu below the box. If you type *April*, for example, you see options for **Filenames** or **Dates**. The Mac wants to know: Are you searching for files named April, or do you want to find files you *created* in April?

When you click to indicate your intention, a *search token* appears in the box—a little oval button that stores the answer to that question, as shown in "Advanced searching" at top.

Here are some of the categories you might see in that little menu:

- **Filenames.** The first option is always **Filenames**, which tells the Mac that, yes, you're looking for files whose names or contents matched what you've typed.

Create search tokens, one for each criterion.

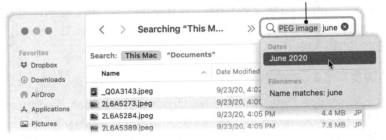

Add a new row for each criterion.

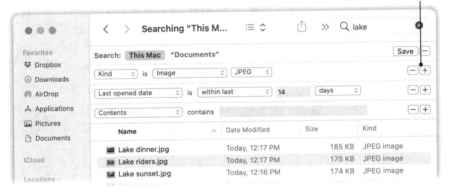

Advanced searching

- **Kind.** If you've typed the word *image*, the Kinds category invites you to specify **JPEG image**, **PNG image**, or **GIF image**. If you've typed *document*, the choices are **PDF document** and **Word document**. And so on.

- **Date.** If you type a date, a month, or a year into the box, the suggestions menu offers corresponding date options. Each one limits the search to files you created, changed, or last opened during that period.

- **Tag.** If what you've typed matches the name of a *tag* you've set up (page 116), the suggestions menu lets you confirm your intention to limit the search to files with those tags. If you type *Important*, for example, the menu might say **Tags:Important**.

If you know exactly what you're looking for, it's perfectly possible to type another search phrase and add a second search token. And a third, and a fourth. In this way, you can build ridiculously complex, specific searches that incorporate many criteria at once.

> **TIP:** There's a similar "search tokens" feature in Mail and Photos.

Of course, the little search box can get pretty crowded once you've created more than, say, one search token. You may find it easier, therefore, to click the + to add a new search refinement, as shown at bottom in the figure "Advanced searching."

How to Save a Search

Once you've set up a Finder window search, you can save it as what Apple calls a *smart folder*. That's a self-updating folder that always contains the files that match a search you've set up in advance.

For example, you could create a smart folder that shows you all the files you worked on this week, no matter what week it is. Or it could contain all the files with the tag *important* that are bigger than 5 MB. You can go about your life, creating and deleting files, and these folders will always be magically up to date.

To create a smart folder, perform a Finder window search as described previously. Once the results appear, click **Save**. Type a name for your smart folder (*This Week* or something) and then hit **Save**.

Saved searches show up here.   *Set up a search; click **Save**.*

Creating a smart folder

Now there's a new icon in your sidebar called **This Week**, at the bottom of the Favorites section. Just click it to produce a window that, in effect, recreates the original search.

Use the time you've saved to start learning a new musical instrument.

CHAPTER SIX

Apps and Docs

S ome apps don't produce new documents. Their entire job is showing something on the screen: games, ebook readers, and video chat apps, for example.

Other apps, though, exist to generate *documents*: files containing your prose, photos, spreadsheets, slideshows, edited videos, and so on. And this is where things can get complicated, because now we're talking about a parent (the app) spewing out dozens or hundreds of children (the documents). You have to track those offspring, organize them, back them up. You have to be able to *find* them again. Sometimes, you have to figure out which children go with which parents.

That's what this chapter is all about: life with apps—and life with documents.

Where to Get Apps

In Part Three, you can learn about the apps that come with every Mac. And, truth be told, those starter apps may be all some people ever need. Word processing, spreadsheets, graphics, web surfing, slideshows, email—it's all there.

But thousands of other apps await you. Some are free or cheap; some cost hundreds of dollars. All of them expand what the Mac can do, and almost all are available from the online Mac App Store.

To look over your options, open the App Store *app*, which is in your Applications folder at this moment, and probably on your Dock as well. As you'll discover right away, the App Store looks something like Amazon.com or any other web store—except everything listed here is an app that you can download immediately.

App categories   The details page for an app

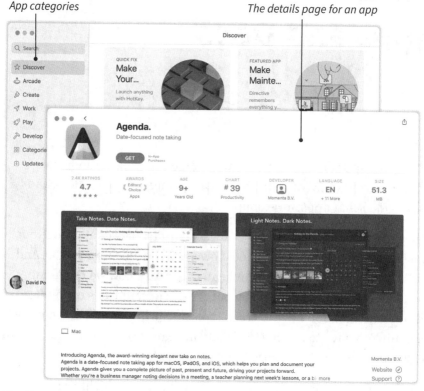

The Mac App Store

If you know what you want, use the search box at top left. Otherwise, you can browse the categories, look over the editor recommendations, and read the reviews left by other people who've bought the apps.

To get a new app, just click its price button and then **BUY APP**. (For free apps, click **GET** and then **INSTALL**.) The Mac downloads and installs it automatically, without requiring passwords, payment, or any other red tape. (Apple already has your payment information as part of your Apple account. How convenient—for both parties.)

When the download is complete, you'll find your new app ready to use—in your Applications folder and on your Launchpad, described next.

Opening an App

Apple wanted to make sure you'll never, ever wind up in the distressing position of not being able to find some app you downloaded. The computer is teeming with different ways to open an app:

- **Start in the Applications folder.** Every app you own sits in the Applications folder. You can get there by choosing **Go→Applications**. Then just double-click the app you want to open.

- **Start at the Launchpad.** Believe it or not, macOS comes with an app whose sole purpose is making all your *other* apps easy to find. It's called Launchpad.

  Because, in theory, you'll be opening Launchpad a lot, there's a *key* dedicated to opening it, right on the top row of your keyboard (⊞). The Launchpad icon is also on your Dock—it looks like nine colorful Chiclets.

Page buttons

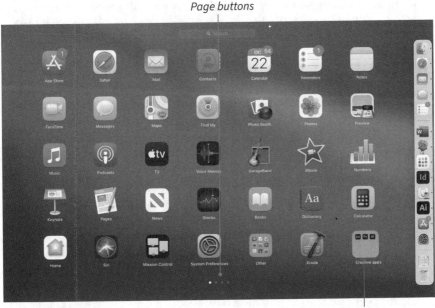

Launchpad   *A "folder"*

And if you have a laptop, you can also open Launchpad by pinching inward with four fingers on the trackpad. It's pretty cool once you get used to it.

If you have a lot of apps, they may fill more than one screenful of Launchpad, as indicated by the little row of dots beneath them. To change "pages," click one of the dots, or swipe horizontally with two fingers on your trackpad, or press ⌘-►. Sooner or later, you'll find a method you like.

To open an app from Launchpad, just click it once.

> **TIP:** Feel free to reorganize the icons on Launchpad by dragging them around. If you drag one on top of another, you create a *folder* that can contain as many apps as you want. (If you've ever used an iPhone or an iPad, this should sound distinctly familiar.)

- **Use the Dock.** Of course, in theory, if there are certain apps you use often, you've installed their icons onto the Dock (page 85). That way, you don't even have to summon Launchpad; one click does the trick.

- **Open a document.** Remember that brilliant parents-and-children analogy for the relationship between apps and documents? The two are forever linked. When you open a document, its parent app opens automatically to show it to you. Double-click a Microsoft Excel document, and Microsoft Excel opens. Double-click a Pages document, and Pages opens.

Switching Apps

You may be good at multitasking in the physical world—or you may not be so good at it—but your Mac is a veritable virtuoso. It can have a huge number of apps open, each playing video or downloading something or crunching numbers, simultaneously.

The problem for you, the human, is that your apps have a front-to-back order, as though they're individual sheets of paper on the screen—and only one can be in front. Wouldn't it be nice if there was some way to unstack them, for the purposes of jumping from one app to another?

There is. In fact, Apple has created a couple of different features for exactly this purpose.

Mission Control

Have a look at the top row of your keyboard. See the strange little ▦ symbol on the F3 key?

When you tap it, all the windows in all the open apps suddenly shrink down and spread out enough that you can see them all simultaneously. Instead of your apps behaving like a stack of papers, they now look like a bunch of Post-it notes, neatly arrayed before you on the desk. The idea is to give you a convenient and instantaneous bird's-eye view of everything open on your Mac. Just click the window you want to bring to the front.

Before: How can you find anything in all this mess?

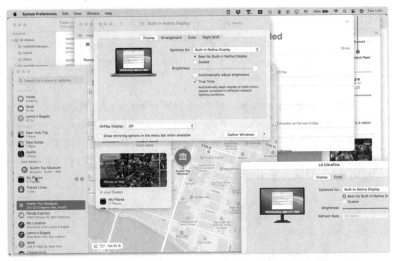

After: Mission Control spreads everything out for your inspection.

Mission Control before and after

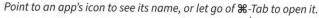
If you decide not to switch windows after all, just press the ⊞ key again to go back to whatever you were doing before.

The App Switcher

Before Mission Control existed, there was the *app switcher*, which is a row of icons showing all your open apps. It's always superimposed in front of whatever else you're doing, for easy access.

To bring up this display, press ⌘-Tab—and then leave your thumb down on the ⌘ key. You can jump to another app just by clicking its icon, or you can use the keyboard: While the ⌘ key is down, tap the Tab key repeatedly. Each icon in turn highlights, left to right; once you've reached the app you want, release both keys. The selected app pops to the front.

Point to an app's icon to see its name, or let go of ⌘-Tab to open it.

The app switcher

More Ways to Hide Windows

Truth is, Mission Control should be all the window-navigation feature you'll ever need. It's quick and reliable, and it shows all your open windows with one keystroke.

Some of the older-fashioned macOS features for hiding the frontmost windows—so you can see what's behind them—are still around, though. Here's a quick summary, just in case you'd like to add them to your Museum of Occasionally Useful Features:

- **Hide the program you're using.** When it might be useful to momentarily hide all the open windows of the program you're using right now, open the application menu. (The application menu is the one right next to the menu, bearing the current program's name, like **Safari** or whatever.) Choose **Hide Safari** (or whatever the app's name is). Immediately, every trace of the app you're using is hidden, so you can see what's behind it.

> **TIP:** You can achieve the same effect by pressing the Option key as you click any visible portion of the desktop or another app's windows.

   To bring your app back, use the ⌘-Tab keystroke, or click the app's icon on your Dock.

- **View the desktop.** Sometimes you might want to hide all windows from all apps so you can get a clear view of the Finder desktop. This feature, Desktop Exposé, comes in handy when, for example, you want to find and open another document that's somewhere on your desktop.

   The trick is to press ⌘-▦. (Yes, it's the Mission Control key with the addition of the ⌘ key.) Like magic, all windows of all apps go flying off the screen to the edges. There they remain until you press the same keystroke again. (Opening an icon in the Finder also brings them back.)

> **TIP:** This is a handy trick when you're composing an email and want to attach a file from the desktop. Press ⌘-▦ to see all open windows make themselves scarce. Find the file you want to attach. Start dragging it—and then, without releasing the mouse button, press ⌘-▦ again to bring the email window back. Without interrupting your drag, bring the cursor directly over the outgoing message window and let go. (Yes, you can attach a file to an email just by dragging its icon from the desktop.)
>
> The whole thing is efficient and magical, and it will impress the heck out of your co-workers.

Quitting and Force-Quitting

You already know that double-clicking an app icon *opens* the app, bringing its windows to the screen and its software into memory. But how do you close it again?

Under normal circumstances, you don't. There's little reason to close the apps you've opened. Your Mac can keep all of them open simultaneously, making it quick and easy to switch among them.

Yes, if you open *enough* apps, you'll begin to notice a slowdown when switching among them. But you won't lose nearly as much time as you would if you quit and then reopened each app in turn.

Quitting an App

Still, it's worth knowing how to exit an app—for example, when it's acting a little glitchy. (Quitting and reopening an app often fixes that kind of thing.)

To do that, open the application menu. Choose **Quit Safari** (or whatever the app's name is). The Mac responds by closing all the app's windows. If you've done any work in a document that you haven't saved yet, the Mac invites you to save your work first.

Quitting an app

Force-Quitting an App

Sooner or later, you'll run into a glitch that's common in computer software: Your app locks up. It's frozen. You can't open its menus, operate its buttons, or even use the **Quit** command. It just doesn't respond at all when you click or type.

In those situations, you'll be happy to know about the Force Quit command. Its purpose is to forcefully exit the frozen program, basically blasting it with a *Star Trek* dematerializer. Once it's gone, you can cheerfully reopen the same app and carry on with your work.

To do that, choose →**Force Quit**. (Or, if you're showing off, press the keyboard shortcut Option-⌘-Esc.) The Force Quit Applications window appears, showing all your open programs. Click the one that's giving you trouble, and then click **Force Quit**. It vanishes from the list of apps—and from your list of problems.

Choose the command.   *Choose the app to kill.*

Force-quitting an app

Feel free to reopen the app and resume what you were doing. (If you were working on a document, however, force-quitting means potentially losing all your work since the last time you used the Save command.)

Save and Open Sheets

Some things haven't changed since the dawn of personal computing, and here's one of them: In any app that creates documents—Word, Excel, Keynote, Pages, TextEdit, Photoshop, and so on—none of the work you do is saved onto your hard drive until you use the **File**→**Save** command. If you don't save your work before you close the app, the work is gone forever.

The first time you use the **Save** command, a weird little floating window appears, attached magnetically to the document window. This is what Apple calls the *Save sheet.* You're supposed to use it to name your new document and choose a folder for storing it. (Most apps propose storing a new document in the Documents folder, just to make life easier on you. As you'll read in a moment, however, you can override that suggestion.)

Unsaved changes

The telltale Close button

The Save Sheet

The Save sheet has two forms: a simplified version and an expanded version.

In its initial, compact form, you can type a name for the file and then glance knowingly at the **Where** menu, which shows where the app proposes storing the new file. You can use this menu to choose a different folder; the choices

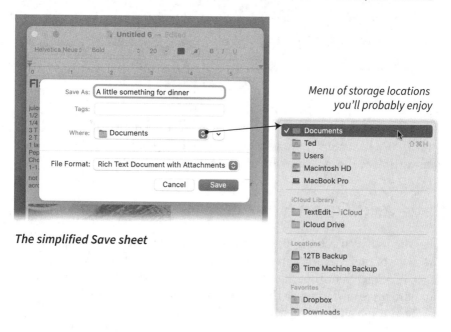

The simplified Save sheet

Menu of storage locations you'll probably enjoy

EMOJI, ACCENTS, AND DIACRITICAL MARKS

Most people don't have much problem figuring out how to use a Mac keyboard. After all, the main set of 26 keys are immediately recognizable to anyone who's made it past kindergarten.

You know what's *not* on any of those keys, though? Emoji, like 💯 and 👀. And symbols, like 👉 and ✴.

You can easily "type" all those expressive characters into most apps, though. The trick is to choose **Edit→Emoji & Symbols**. A special palette appears, containing row after scrolling row of emoji and symbols that you can simply double-click to enter into your document. (In the Messages app, the one where you're most likely to use emoji, there's an easier way to open the palette; see page 228.)

At first, the palette shows emoji, organized into eight categories (click one of the icons at the bottom), plus a Frequently Used category (the ⏱). Don't miss the fact that you can scroll down within any category's "page."

But if you click the 😀 icon, the panel changes to symbols, organized into nine categories like Bullets/Stars, Math Symbols, Pictographs, and so on. Scroll down through each category page to find just the symbol, and then double-click it to insert it.

Actually, there's one more group of typed characters that doesn't appear on your keyboard: accents and diacritical marks, as in the words "naïve" and "flambé."

To type one of those, press the key for the *letter* you want affected by the symbol—and wait a full second. Like magic, a little pop-up panel of markings that you can use on that letter appears. For example, when you press the E key, you get a choice of **è, é, ê, ë, ē, ė, or ę.** You can click the one you want to "type," or, if you'd rather not lift your hands off the keyboard, you can type the number that appears in light gray beneath the character you want.

Now anyone can be a polyglot. ✦

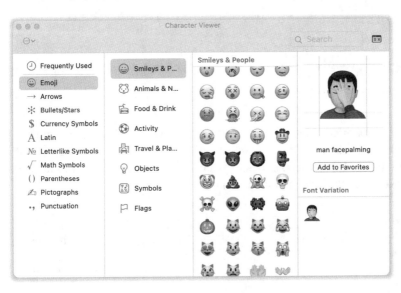

include all your drives, your iCloud Drive, folders you've used recently, and any folders in your sidebar Favorites list (page 94).

Once you've indicated where you want to save the file, click **Save**.

But if you click the ˅ button next to the **Where** pop-up menu (or press ⌘-=), the window expands into what's basically a full-blown Finder window. Now you have a choice of views (list view, icon view, column view), the option to search or sort your files, and even the standard Finder sidebar at the left side, offering you access to other disks and locations.

Click to switch between the simple and expanded forms of this box.

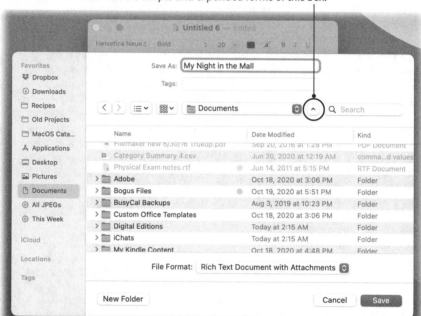

The expanded Save sheet

In this expanded view, you can create a new folder to receive your newly hatched document by clicking **New Folder**. In some apps, there's a **Format** menu here, too, so you can specify what kind of document you want to create (a text file, a Word document, or whatever).

> **TIP:** You can rename an existing file, or even delete it, right in this Save As window. Right-click or two-finger click its name; from the shortcut menu, choose **Rename** or **Move to Trash**. Revel in your power.

The Open Sheet

Sometimes, when you're already in an app, you may want to open a different document without having to return to the Finder. That's the purpose of the File→Open command. It produces a sheet that looks almost exactly like the expanded Save sheet described already.

Your job here: Navigate to the folder that contains the document you want—and, once you've found it, double-click to open it.

Truth is, you may not see much of the Open sheet in your Mac travels. If your goal is to open some existing document, it's usually just easier to hop back to the familiar desktop (⌘-🔲) to hunt down the document.

Printing

So weird—1980 came and went, but the paperless office never did arrive. It is true, though, that people print documents a lot less often than they used to.

Connecting to a New Printer

Most personal printers—inexpensive ones for home use—plug into your Mac's USB jack. Hook yours up and turn it on.

To complete the setup, you have to introduce the Mac to its new printer. The quickest way: Open something you want to print—an email or a web page, for example. Choose File→Print. In the Print dialog box that appears, open the Printer pop-up menu and choose the new printer's name.

> **NOTE:** In some small-office situations, the printer's name may be hiding in the Nearby Printers submenu. And if you *still* don't see your printer's name, open **System Preferences→Printers & Scanners**. Click **+**; in the resulting box, click the printer's name, and click **Add**.
>
> And in *big*-office situations—well, a highly paid network geek is supposed to set your Mac up to use the company's printers.

That's it. That's the entire user guide for connecting a printer.

Making the Printout

Now comes the moment of truth: making an actual printout.

With the printable document on the screen before you, choose File→Print. The Print box that appears is, like so many Mac dialog boxes, an

Expand-O-Box. Microsoft Office apps, like Word and Excel, use a different design, but in most other apps it has two forms:

- **A compact, simplified version.** This box shows a preview of your printout-to-be. It has just a few options. They're what most people need

Simplified Print box

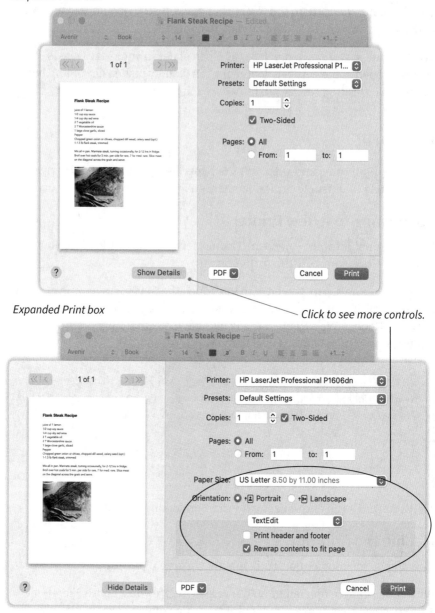

Expanded Print box

Click to see more controls.

The Print box, little and big

to adjust most of the time: the number of copies, the range of pages to print, and the printer you intend to use.

> **TIP:** You don't have to fill in both the From and To boxes; save precious finger movement! If you leave the first box blank, the Mac assumes you mean "From page 1." If you leave the second box blank, it assumes you mean "To the end." For example, if you need only the first two pages of the document, leave the first box blank and type *2* into the second box.

- **An expanded, detailed option** appears when you click **Show Details.** This version offers all kinds of controls that are unique to your printer: the paper size and orientation, for example, and—in the pop-up menu— options like **Layout** (how many pages you want to print, at reduced size, on each physical sheet), **Media** (what paper tray you want to use, if your printer offers more than one), **PDF**, and so on.

Choose the options you want, and then hit **Print** to begin the printing.

The Printer Queue Window

During the printing process, a weird little printer *app* opens, named for your printer; it appears both in your Dock and in your app switcher (page 136).

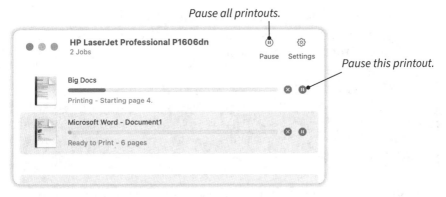

The printer queue window

The thing is, the app immediately quits when the printing is finished. In other words, if you're printing just a page or two, you may have to look fast to see the little printer app.

That's OK, because the primary beneficiaries of this window are people who share a printer with a bunch of other people, print a lot of things at once, or have access to several printers on a network.

The printer queue window is the software dashboard for the printing process. It lists all the documents that are lined up to print, in order. At this point, you can remove a document from the schedule to be printed (click it and then click the ⊗ button), rearrange the printouts (drag them up or down), or pause all printing (click ❙❙, and then click ⊙ when you're ready to resume). Pausing might be useful when you need to fill a paper tray, for example.

The Settings button (⚙) lets you rename the printer or check its ink levels.

Eleven Keystrokes Worth Learning

At the beginning of your Mac career, you may take comfort in knowing that every command you'd realistically ever want to use, in every app, is listed in the menus at the top of the screen. But, over time, you may come to value the efficiency of using keyboard shortcuts. Once you've got a few under your belt, you'll discover that they save you time (because you don't need to sacrifice a hand to move the mouse to the menu) and fumbling (because you don't need to know where the command *is* in the menus).

Add up all the commands in all the menus of all the apps, and we're talking about thousands of keystroke combos. But if you learn only these 11, you'll gain a lot of speed and power without a lot of work:

- **Close window: ⌘-W.** An essential. Why aim for that tiny red dot at upper left when you can swiftly dispatch the thing with a quick flick on the keyboard? (Works in almost all apps.)

> **TIP:** In most apps that produce documents, you're supposed to *save your work* periodically, preserving the latest version on your hard drive, by choosing **File→Save** (⌘-S). But what if you try to close a window when you've done some work and *haven't* saved the latest changes? Fortunately, the Mac will ask if you're sure before it actually closes anything.

- **Undo: ⌘-Z.** Most software companies are aware that you're human. Therefore, you make mistakes. Therefore, you'll value the Undo command. It takes back the most recent step you took, whether it's putting icons into the Trash, deleting some text in Notes, pasting an image in Photoshop, or whatever. Some apps, in fact, let you undo *multiple* steps, Undo-Undo-Undo, going all the way back to the moment you first opened the document.

Obviously, having to open a menu (**Edit→Undo**) 39 times in a row would drive you mad. Fortunately, ⌘-Z does the same job.

HOMEMADE KEYBOARD SHORTCUTS

There are two kinds of people in the world: those who thrive on the efficiency of using keyboard shortcuts to do everything, and those who prefer to use the mouse (and who keep saying, while reading this book: "What is this guy's *obsession* with listing keystroke alternatives to perfectly simple tasks?").

If you're in the second category, you can safely skip this box.

If you're in the first, however, you may be delighted by this news: You can actually *create or change* any key combination for any menu in any app.

If the command for **Check Spelling** in your favorite word processor comes from the factory as Fn-Shift-Option-⌘-F11, and you think that's way too complicated, you can change it to Control-C.

If the command for **Rotate** in your favorite photo-management program doesn't even *have* a keyboard shortcut, you can make one up—for example, ⌘-R.

To begin, note the *exact* wording of the menu command in question. (The name of the *menu* it's in doesn't matter.)

Open **System Preferences→Keyboard**. Click the **Shortcuts** tab; in the list at the left side, click **App Shortcuts**. Click the **+** button below the list.

In the little "sheet" that appears, use the **Applications** pop-up menu to choose the name of the app. In the **Menu**

Title box, type the name of the menu command you wrote down. Capitalization matters. Sometimes, a menu command is followed by an ellipsis (…). You produce that symbol by pressing Option-semicolon (;).

Finally, in the **Keyboard Shortcut** box, press the keyboard combination you'd like. In most apps, many menu commands are already assigned to letter combinations involving the ⌘ key: ⌘-P for Print, ⌘-C for Copy, ⌘-Z for Undo, and so on. But the other modifier keys—like Option, Shift, and Control—are generally available for use with letter keys, and *combinations* of modifier keys are generally wide open.

When you've come up with a keyboard shortcut you can remember, click **Add**. Pop back into the app, where you'll see that your new keystroke actually appears next to the corresponding menu command.

Give it a try, and marvel at your own ingenuity. ★

- **Cancel that menu, dialog box, or mode: Esc.** The Escape key (Esc) always means "Back out of this." If you click to open a menu, Esc closes it without choosing any command. If you've opened a dialog box (to print something, for example), Esc closes it without doing anything. If you're in YouTube and you've made the window full screen, Esc exits full-screen mode. If you've started to drag an icon in the Finder and you change your mind, Esc cancels the drag and puts the icon back where it started.

- **Switch windows in the same app: ⌘-~.** You already know about the ⌘-Tab keystroke for switching among open apps (page 136). But you can also press ⌘-~ to hop to a different open window in the *current app*. (That's the tilde key, the wavy squiggle to the left of the number 1 key.) For example, maybe you have a bunch of Word documents open, or a bunch of browser windows, or a bunch of Photoshop documents. Each time you press ⌘-~, the Mac presents the next window in whatever app you're using now.

- **Copy, Cut, and Paste: ⌘-C, -X, and -V.** Copying (or cutting) text, graphics, numbers, bits of video or audio, and even desktop icons is a cornerstone of computing—and so, of course, these keystrokes should become part of your muscle memory. The mnemonic scheme isn't bad: The three keys are adjacent on the keyboard. C for Copy, X for Cut (like a pair of scissors), and V for Paste (because it looks like an "insert here" arrow).

- **Bold, Italic, Underline: ⌘-B, -I, and -U.** These keystrokes work in any app where you can add these styles to text: Mail, Pages, Keynote, Word, Notes, Stickies, and so on. (Some apps, including Messages and Reminders, show only plain text and can't do bold, italic, or underline.)

  To apply one of these styles to text you've already typed, select it first and then hit the keystroke. But as you're typing along, you can also press the keystroke *before* you type, thereby turning on the boldface (or italic or underline), and then press it a second time *after* typing the word or phrase to turn it off … and keep right on typing.

- **Spotlight: ⌘-space.** Use it whenever you need to find something, either on your Mac or online. Chapter 5 has the details.

CHAPTER SEVEN

All About iCloud

A pple controls both ends of the connection between its internet services and your Mac, iPhone, or iPad. That struck somebody at the company as an amazing opportunity.

What if it could use the internet to synchronize the contents of all your gadgets? You could add somebody to your address book on the Mac, and it would also show up on the iPhone and your iPad. You could take a picture with the phone, and it would appear on your iPad and your Mac. You could add a bookmark to a web page on the iPad, and it would appear in the Favorites list on your Mac and iPhone. You could create a calendar appointment on one gadget, and it would appear instantly on all your others. And so on.

All that may sound like a big "duh" to you, the modern techno-citizen. But before iCloud came along, those were difficult scenarios. When you wanted to look at a photo, consult an email, or look up a phone number, you'd have to remember which device it was on.

In any case, that's only the beginning of the suite of free services that Apple calls iCloud. Yes, it does an amazing job of synchronizing your notes, reminders, appointments, email, photos, bookmarks, voice memos, passwords, news and stocks preferences, and other data across all your Apple gadgets. But iCloud also includes the iCloud Drive (a backup hard drive in the sky), a free email account, Find My iPhone, the ability to share stuff you buy from Apple's online stores (movies, music, books, apps) with family members, and much more.

The iCloud Account (Apple ID)

Like most self-respecting global tech behemoths, Apple requires that you sign up for a free account before accessing its universe of free services. Your name and password are called your Apple ID; you'll be asked to supply it many times in the coming years. Among other things, your credit card information and contact information are part of your Apple ID, which can save you all kinds of time and tedium reentering this information online.

The very first time you install macOS Big Sur or turn on a new Mac, you're invited to sign up for an Apple ID. If you somehow missed that opportunity, you can make up for your tragic oversight by opening **System Preferences→Apple ID** and clicking **Create Apple ID**. You're asked for your name and birthdate.

If you have an existing email address, it can become your new Apple ID—but you can also click **Get a free iCloud email address**, which is a useful option. Now you'll have *two* email addresses—your old one and your new "@icloud.com" one. That way you can supply one address whenever it's requested by websites; eventually, it will be harvested by the scum of the internet and become targeted with spam, ads, and hacking attempts. You can reserve the other email address exclusively for private communications, confident that it will remain pure.

> **NOTE:** The centralized hub of Apple ID info is in **System Preferences→Apple ID**. Here, on panels called things like **Overview**, **Password & Security**, and **Payment & Shipping**, you can examine or edit all the aspects of your Apple ID, change your password, edit your credit cards, and so on.

Synchronized Data

That business of using the internet as a coordination channel between all your Apple machines is truly a blessing—and it's part of the velvet handcuffs that keep people in the Apple ecosystem. After all: If you were to switch to some other kind of phone or computer, you'd have to deal with the hassle of recreating your calendar, address book, notes, reminders, bookmarks, voice recordings, photos, and so on.

In any case, the headquarters for data synchronization is in **System Preferences→Apple ID**. If you click **iCloud** in the sidebar, you see a list of the kinds of data the service can synchronize across your gadgets, each with its own on/off checkbox. For example:

TWO-FACTOR AUTHENTICATION AND YOU

Let's face it, people. Passwords are a bust.

Not just because they're a pain to memorize and to type—but because they're actually not very secure. How many times a year do we hear about data breaches, where a bad guy makes off with millions of customer names and passwords?

Security experts have come up with an ingenious way to keep your account protected even if somebody steals your password. This system is pretty great; it has shut down the kind of data thefts that put naked pictures of Hollywood stars online in the great iCloud hack of 2014.

Really, the only unsuccessful part of it is the name: *two-factor authentication*, a term nobody can remember and nobody understands.

In fact, it means just what it says: When you try to access your iCloud account on a new machine, you're going to need more than just your password. You also need a second factor: a six-digit code that Apple displays on all your existing Apple gadgets. If it's actually some Russian hacker pretending to be you on *his* Mac, he'll never get that code, and he'll be stopped at the gates.

Lots of Apple features don't even *work* unless you've turned on this security layer. The setup software for macOS Big Sur pushes you pretty hard to set it up.

If you haven't already, open **System Preferences→Apple ID→Password & Security**. In the Two-Factor Authentication section, click **Turn On**. The Mac asks for your phone number, which it uses to send you a verification code.

Now that "2FA" is turned on, here's what will happen the first time you try to use a new Apple device, or the first time you try

to access your iCloud account in a new web browser.

On all your Apple gadgets, you see a map and the words "Your Apple ID is being used to sign in to a new device." If you don't recognize the location, somebody is trying to hack you, and you can click **Don't Allow** to slam the door.

If you click **Allow**, though, Apple sends a one-time, six-digit code to every one of the Apple gadgets it already knows you own. Type it into the awaiting six-digit boxes. (If you don't have more than one Apple device, Apple can send the six-digit code to your other phone number.)

You've just added a new "trusted device" to the list of machines that Apple knows are yours. They show up in a list at the left side of the **System Preferences→Apple ID→Password & Security** screen.

This is a one-time deal for each new web browser or Apple product you use. After that, the Apple mother ship recognizes that it's you—and thanks you for your patronage. ★

- **Mail, Contacts, Calendars, Reminders, Notes.** These apps exist in almost identical forms on Macs, iPhones, and iPads—and these checkboxes ensure that their contents will be identical, too.

> **NOTE:** The Voice Memos app also exists on every device, and can also synchronize its data (your audio recordings). For reasons known only to Apple, though, it doesn't have a checkbox on/off switch on this screen. Instead, in **System Preferences→Apple ID**, click **Options** next to **iCloud Drive**. *That's* where the **Voice Memos** switch sits.

- **Safari is Apple's web browser,** and this checkmark ensures that your bookmarks (Favorites) and Reading List (page 206) will appear identically across gadgets.

- **Keychain** means "memorized passwords." Not just for websites, but also for logging into other Macs or PCs on the network. Passwords are a bane of our modern existence, so it would be hard to imagine why you wouldn't want these memorized and synchronized between your Apple gadgets.

- **News, Stocks, Home, Siri.** These checkboxes sync your settings and preferences. For example, once you set up your stock portfolio in Stocks on the Mac, it will show up identically in Stocks on the iPhone. Same thing with your news-publication preferences, your home-automation setups, and what Siri has learned about the way you speak.

- **Photos.** If you turn off this checkbox, you won't be able to use the iCloud Photos feature described on page 286. You won't be able to share albums of photos with other people, either. In other words, this is a master switch for any photos leaving your Mac to travel on the internet.

Continuity

If you own both an iPhone and the Mac, this one's for you. The suite of features Apple calls Continuity involve using the iPhone as an accessory to the Mac. Now you can use the Mac as a speakerphone; behind the scenes, your iPhone does the dialing. You can use your Mac to write text messages; your iPhone is actually sending them. You can write your signature on the iPhone with your finger, and drop it into a contract on the Mac. You can copy something from a web page on the Mac, and paste it into an app on the iPhone. You get the idea.

Calling from the Mac

Even if your iPhone is somewhere else in the house, even if it's asleep and locked, it can serve as the cellular antenna for your Mac. That's right: The Mac is now a speakerphone.

When a call comes in, your Mac rings and displays a notification. Click **Accept** and say hello.

And when you want to place a call, click the ☎ next to any phone number in Contacts.

THE APPLE WATCH LOGIN TRICK

If you're the lucky owner of an Apple Watch, Continuity provides a really slick trick: You can log into your Mac without a password or fingerprint. The Watch's proximity to the Mac can serve as a wireless key.

To set this up, open **System Preferences→Security & Privacy→ General**. Turn on **Allow your Apple Watch to unlock your Mac**.

You must also have two-factor authentication turned on for your iCloud account, as described on page 151. (Apple is especially finicky about security for this Apple Watch feature. It would really like to avoid headlines about passing strangers unlocking your laptop with their Apple Watches.)

From now on, just wake your Mac, marvel as "Unlocked with Apple watch" appears briefly on the screen, and you're in.

In fact, there's another way to use this feature. Whenever you're supposed to enter your password or use your fingerprint—for example, to change the time zone in System Preferences, view your passwords in Safari, or unlock your protected pages in Notes—you can just double-click the Watch's side button instead. It's quick, it's easy, and it seems like magic. ★

Texting from the Mac

You can also send and receive text messages from your Mac, which has a glorious full-size keyboard. It's much easier to type on than glass.

To get this set up, start on the iPhone. Choose **Settings→Messages**, and turn on **Text Message Forwarding**.

Now, on the Mac, open Messages. The first time, it displays a numeric code, which you're supposed to type into the corresponding box on your phone. All of this is a security measure, designed to ensure that only *your* Mac can get your iPhone's messages.

From now on, you can use Messages to send standard texts to any cellphone number. (You can enter normal cellphone numbers in Messages—your correspondents don't have to be members of the Apple cult.)

You can also right-click or two-finger click any highlighted phone number in apps like Contacts or Notes—and, from the shortcut menu, choose **Message 800-555-1212** (or whatever the number is) to commence texting.

Your Mac can also *receive* text messages. They appear as standard notification bubbles, complete with the option to reply. (See Chapter 10 for more on using Messages.)

Continuity Camera

Almost every Mac model comes with a built-in camera. Unfortunately, it's not high-quality. It doesn't do well in low light. It doesn't have a flash or a zoom. And it's not very easy to position. You can't hold it like a phone, making it higher or lower, angling it this way or that.

Fortunately, you don't care. Continuity Camera lets you use your iPhone to take a picture, which instantly appears on the Mac.

This feature works in most of Apple's built-in apps: Mail, Messages, Notes, Preview, Pages, Keynote, and TextEdit, for example. To try it out, right-click or two-finger click a blank spot in the document. From the shortcut menu, choose **Take Photo**.

> **TIP:** This shortcut menu also offers a **Scan Documents** command, which opens the iPhone's document-scanning mode, complete with some kind of magic that automatically straightens and crops whatever piece of paper you're scanning. Finally, the menu offers **Add Sketch**, which lets you draw a freehand sketch—or perhaps it should be called a *freefinger* sketch—on your phone.

*1. Right-click.   2. Click **Take Photo.**   3. On the phone, take the shot and tap **Use Photo.***

Continuity Camera

Freakishly enough, the camera app automatically opens on your iPhone or iPad. Frame up the shot, snap it, and if you like the result, tap **Use Photo.** A second later, the photo appears on your Mac, pasted right into the document.

Continuity Markup

As you'll learn on page 298, the Preview app seems to offer three different ways for you to sign a document electronically. In fact, though, there's a fourth way: You can sign or annotate the document on your phone, and transfer the markings by Continuity to the Mac.

Throughout the Mac universe—in Mail, Notes, TextEdit, Photos, Quick Look, and so on—you run across the markup tool (Ⓐ). It's a handy little drawing toolkit with various markers and colors you can use to annotate a graphic or PDF file. Which is great—if you can endure the clumsiness of drawing or writing with the mouse or trackpad.

Instead, try this. In the Finder, click a PDF or graphic document's icon and then press the space bar to open Quick Look (page 103). Click the Markup tool (Ⓐ), and then click the ⬚ at the right end of the toolbar. The image or document you're editing magically shows up on your phone or iPad, complete with the same set of markup tools. Add your signature, cross out certain phrases, make your proofing marks, and exploit the magic of having a touch-screen. As you draw, in real time, those markings fly wirelessly through the air to the Mac version of the image. Tap **Done** to seal the deal.

Continuity Markup

Continuity Clipboard

This feature lets you copy text or graphics on the Mac and paste them on the iPhone or iPad. Or go the other way. There's nothing to it; there's no special button or toolbar.

On the iPhone or iPad, copy something. Now, on your Mac, choose **Edit→Paste**. Incredibly, whatever you copy on the iOS device appears on the Mac.

> **NOTE:** You have two minutes to do the pasting. After two minutes, whatever used to be on the Mac's own invisible Clipboard reappears.

Family Sharing

In the olden days, being a member of an Apple-device family could get a little complicated. Every time your kids wanted to buy some app, movie, or book, they'd have to bug you for your credit card. Also, it made no sense that once your spouse had paid $20 for a movie, you would have to pay $20 again for the same movie, because you had a different Apple account. You're married, for heaven's sake!

Family Sharing solves all that. You identify up to six people as belonging to the same family, and just look at all you get:

- **Shared purchases.** All of you can share the same unified stash of music, books, movies, TV episodes, and apps.

- **Shared credit card.** Everyone can make charges on the same master credit card. It's not exactly Kids Gone Wild; when one of your kids tries to buy something, a notification appears on your screen. You can approve or decline the purchase on the spot.

Permission notification

- **Shared storage.** All of you can share an iCloud storage plan. For example, if you're paying $3 a month for 200 GB of storage, all family members can share it for use with iCloud Drive, iPhone backup, and so on.

- **Shared subscriptions.** Everybody can enjoy the family-plan versions of Apple's monthly subscription services, like Apple Music, Apple Arcade, Apple TV+, News+, and Apple One (which is a single master subscription to all the others).

- **Find one another.** You can see where your family members are on a map (with their permission, of course), using the Find My app (page 259) or by logging into iCloud.com.

- **Monitor your kids' addictions.** You can see how much time they're spending on their Apple gadgets and what they're doing.

- **Find lost gadgets.** You can also find one another's lost Apple gadgets.

- **Shared appointments, albums, and reminders.** This is a little thing, but handy: When you turn on Family Sharing, a new category called Family appears in Photos, Calendar, and Reminders. This Family category is configured so everybody in the group has access to it.

Setting Up Family Sharing

To set up Family Sharing, open **System Preferences→Family Sharing.** Walk your way through the screens that explain how this is all going to work. Confirm that you will be the organizer, the one who has the wisdom and experience to oversee Family Sharing for everybody—and to pay for what they buy. Declaring yourself the head of the household also means specifying a credit card for all their charges. You can indicate whether or not you want *your* purchases and physical location shared with your family.

At this point, you can add up to five more people to the family. Click **Add Family Member** to begin; instructions will guide you through supplying (or creating) this person's iCloud account.

> **NOTE:** Ordinarily, you must be over 13 to have an iCloud account. But the Family Sharing feature makes an exception. You, the parent, can create a special under-13 iCloud account just for this purpose by clicking **Create an Apple ID for a child who doesn't have an account**.
>
> Don't fool around with creating Child accounts just for fun, though. Once you've created such an account, you can't delete it until that "kid" is 13.

Managing Family Sharing

Once you've got everybody set up, you can open **System Preferences→Family Sharing,** click **Family,** and look over the smiling faces (or boring first-letter initial avatars) of your immediate relatives. For each person, you can click **Details** to make changes like these:

- **Ask To Buy.** This option, available on children's accounts, means that whenever your kid tries to buy something from an Apple store, *you* will be notified, and you'll have to give your permission. If you trust the kid, on the other hand, you can turn this option off.

- **Parent/Guardian.** This option, available for adult accounts, lets you designate somebody else to be an approver of the kids' purchase attempts. Your spouse, for example.

- **Remove from Family Sharing.** When somebody leaves the family group—someone leaves the nest, gets disowned, gets divorced—you can remove them from the Family Sharing group. You've just opened up one of those six precious slots for somebody else.

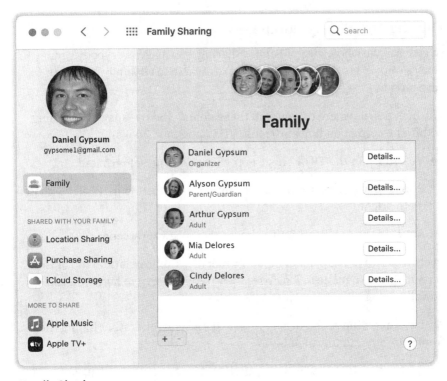

Family Sharing

In the sidebar of this System Preferences panel, you can see a list of the individual Family Sharing features: **Location Sharing, Purchase Sharing, iCloud Storage**, and so on. These are the on/off switches for these features. Some, like **Location Sharing** and **Screen Time**, let you turn off the feature for each family member individually. (For **Screen Time**, you're allowed to spy on family members' activities only if they're under 18.)

If you're *not* the family organizer, you can only see, not edit, most of the settings here. But these two things you can change:

- **Purchase Sharing.** You can stop sharing your purchases with the other family members. Maybe you really don't need your siblings knowing that you bought an app called *AcneKill 2000: Ten Easy Steps to Clearer Skin.*

- **Location sharing.** You can choose which family members are allowed to track you.

Finding and Hiding Purchases

In general, the whole idea of Family Sharing is to make everybody's purchases (music, movies, TV, apps, books) available to all family members, to save money.

The question is, where do you find these items? And the answer is "hiding within the corresponding app (Music, TV, App Store, Books)":

- **TV, Music.** In the TV or Music app, choose Account→Purchased.

- **Books.** In the Books app, choose Store→Book Store Home. Click Purchased on the right.

- **Apps.** In the App Store app, choose Store→View My Account.

In each case, a pop-up menu of family members appears at the top of the screen. Choose a name to look over their purchases; from here, you can download them directly to your Mac.

Ah, but what if the problem isn't *finding* purchases, but rather hiding them from your other family members?

All you have to do is to view your purchases in the corresponding app, as just described. But this time, point to the thing you want to hide, click ⊗, and then click Hide. (In the App Store app, click your name at lower left, and then point to the app you want to hide. From the ●, choose Hide Purchase.) You still own it, and you can still use it, but nobody *else* sees that you have it.

Everything Else

Some of iCloud's best features don't require you to learn anything or click anywhere: They're just *there*. For example:

- **iCloud Drive.** The iCloud Drive, as you may remember from page 105, is Apple's version of Dropbox: a simulated hard drive, apparently on your screen but actually out there on the internet, that all your Apple gadgets can access.

iCloud Drive is therefore an extremely useful place for storing files and folders that you might want to work on from different places. It also allows you to share files and folders with other people over the internet—files that are much too big to send as email attachments.

- **Managing Purchases.** Apple's online stores generate millions of dollars in sales of music, movies, TV shows, books, and apps. But here's the generous part: You're allowed to download anything you've purchased to any Mac, iPhone, or iPad that's signed in with your iCloud account. Or re-download them, even if it's been years since the purchase.

  Of course, iCloud also remembers where you stopped reading each book or watching each video.

- **Automatic backup.** You can set up your iPhone or iPad to back itself up automatically to your iCloud account, wirelessly and invisibly.

PART THREE

macOS Piece by Piece

CHAPTER EIGHT

Mail

The invention of the smartphone may have led to an explosion of communications channels—FaceTime, Messages, Facebook Messenger, WhatsApp, WeChat, and so on. But no app has attained the universal reach, the old-school literacy, or the official paper trail-ness of good old email.

The Mac's email app is called Mail. And it's so complete and so well-integrated with the rest of macOS that it's hard to find anybody who uses any *other* email app.

Plenty of people do access their email directly on the web, using Gmail accounts, Yahoo Mail, Microsoft Live Mail, or even AOL Mail. But Mail is worth considering as a front end for all those services, too. It consolidates all your email accounts in one place, gives you a more complete feature list, and lets you duck out of the advertising that comes with using the web.

Mail Setup

The first time you open the Mail app, it asks what kind of email service you have: **iCloud, Gmail, Yahoo, AOL,** or **Microsoft Exchange**. If indeed you use one of these free services, then setting up your email is as simple as supplying your email address and password.

> **NOTE:** If it ever comes time to add another account, you can return to this Add Account screen by choosing **Mail→Add Account**. Actually, **System Preferences→ Internet Accounts** brings you to that setup screen, too.

If, on the other hand, you get your email from some oddball service not listed here, click **Other Mail Account**. You'll have to supply the login specifics manually, which may include your name, password, and mail-server details.

Mail account setup

Once everything is set up, Mail goes online and starts retrieving mail. In most cases, you'll even get all your *old* mail—read and sent—so that what you see here perfectly matches what you would see on your email service's website (for example, Gmail.com).

Mail Basics

Mail automatically checks for new messages every few minutes. You can also force it to check by clicking ✉ on the toolbar, or by choosing **Mailbox→Get All New Mail**.

The Mailboxes List

The Mail sidebar lists all the folders that contain your email. You get a separate heading for each email account you use (**iCloud**, **Gmail**, and so on). You can collapse or expand the folders within an account by pointing to its name and clicking ❯ or ⌄.

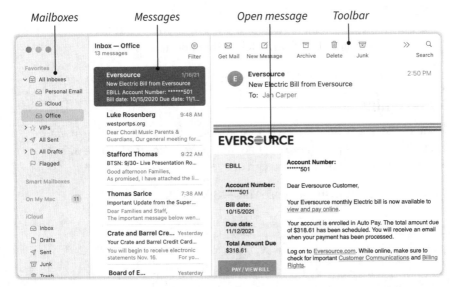

Mailboxes   Messages   Open message   Toolbar

The Mail app's layout

Within each account heading, you'll generally find folders like these:

- **Inbox** is mail you've received. If you have more than one email account, click **All inboxes** to see all the mail, consolidated from all the accounts. Alternatively, you can click the flippy arrow to view a separate inbox folder for each one.

- **VIPs** holds mail from the most important people in your life—those you've designated as VIPs (page 185). (This folder doesn't appear until you've designated some VIPs.)

- **Flagged**. As you work through your email messages, you can apply colorful little flags to them, meaning whatever you want them to mean. This folder rounds up all the messages, from all your accounts, bearing each flag color. (The **Flagged** folder doesn't appear unless you have, in fact, flagged some messages.)

- **Outbox** lists messages you've written but haven't yet sent. It doesn't appear if there aren't any waiting messages to send.

- **Drafts** holds messages you've started writing but aren't ready to send.

- **Sent** contains copies of all the messages you've sent.

- **Trash** holds your deleted messages. As in the regular Mac Trash, this one is only a waiting room for deletion; messages you've put here remain here until you *empty* the trash and delete them permanently.

- **Junk** appears when you use Mail's spam filter (page 186).

Redesigning Mail

If you're like many people, you'll be spending many hours staring at your email. Mercifully, Apple has designed Mail to be infinitely flexible in its layout. Here's some of what you can do:

- **Hide the folders list.** Choose View→Hide Mailbox List.

> **TIP:** Hiding the folders list gives you more screen space to work with, but it also ... hides the folders list. How are you supposed to access the folders you need often?
>
> Easy: Install them on the Favorites bar. Once you've made it appear (**View→Show Favorites Bar**), you can drag any folder onto the gray strip just above the messages list. There it becomes a button—but it still behaves as though it's a folder. You can click it to see what's inside, or drag a message onto it to file it there. (To remove a folder from the Favorites bar, just drag it downward.)

The existing folder buttons slide apart to make room.

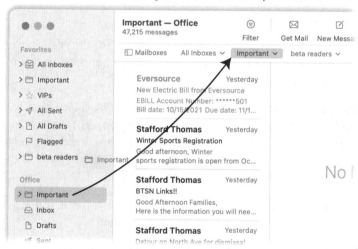

Bookmarks and folders on the Favorites bar

- **Sort the messages.** Using the View→Sort By options, you can sort the list of messages by **Date, Attachment, Unread,** and so on. The bottom two commands—**Oldest Message on Top** or **Newest Message on Top**—control the sorting direction.

- **Add button labels.** The toolbar buttons, like ✉ and 🗑, start out without any text to identify them. But to help you remember what they do, try this: Right-click or two-finger click the toolbar. From the pop-up menu, choose **Icon and Text** (or even **Text Only**).

- **Change the type size for the folders list.** You'd never guess where to find this setting: in **System Preferences→General→Sidebar icon size**. Weirdly, that pop-up menu governs the font size for both the Finder sidebar *and* for the Mail folders list.

- **Adjust the columns.** You know how Mail is designed as three columns—folders, messages, and message body? You can drag the divider line—the right edge—of any of these pieces to adjust their widths.

> **TIP:** If you double-click the right edge of the folders list or the messages list, Mail makes it snap to what it considers the perfect width.

- **Redesign the messages list.** The messages list—the middle column—gives you a little preview of what's inside the message: the sender's name, the subject line, and the first few lines of the message. But using the **View** menu, you can add other bits of information to this list, including **Message Size** or **Contact Photo** (a little picture of the sender, if indeed you have a photo of her in Contacts).

You can also specify how many lines of the preview appear in this list—from zero to five lines of the message body. Visit the **Mail→Preferences→Viewing→List Preview** pop-up menu.

> **TIP:** Suppose you've scrolled way down into the message list. Instead of painstakingly scrolling back up to the top, here's a shortcut for you: Just click the **Sort by** bar above the list. You teleport straight to the top.

The Mac's two insta-workspace-overhaul features are available in Mail, too, and they work especially well in this app:

- **Full Screen mode.** Click the green ● dot in the upper-left corner to make the Mail window fill your monitor.

- **Dark mode.** Dark mode looks really cool, and can be refreshing on the eyes. But in dark mode, type generally appears as white writing against black, which can get weird when you're reading a lot of email.

Fortunately, in **Mail→Preferences→Viewing**, you can turn off **Use dark backgrounds for messages**. Mail continues to display the menus, lists, and panels in dark gray, but the actual message bodies are white.

Composing Messages

To write a new email message, click ⎘ in the toolbar (or choose **File→New Message**).

Now just fill in the various pieces of the New Message window:

- **To.** Enter the recipient's email address. As you type, Mail automatically fills in the rest of the email address to save you time and typos. It draws its guesses from the list of people in your Contacts app, plus recent Mail correspondents.

> **TIP:** If Mail keeps proposing someone you don't really communicate with, you can remove that address from its memory. Choose **Window→Previous Recipients**, click the address, and then click **Remove From List**.

  If you're sending to more than one person, type a comma after each address.

- **Cc.** On the **Cc** line ("carbon copy"), enter the addresses of people who should get a copy of this message, even though they're not the primary recipients.

- **Bcc.** Any addressee you enter here, on the **Bcc** (blind carbon copy) line, gets a *secret* copy of the message. The other recipients don't see any evidence of it. (If you don't already see this line, choose **View→Bcc Address Field**.)

> **TIP:** As soon as Mail recognizes that you've entered a complete email address, that person's name becomes a shaded *button*. Each one has a little menu (⌄) that lists useful commands like **Edit Address**, **Copy Address**, and **Show Contact Card**), as well as this person's other email addresses. You can also drag these buttons—between the **To** and **Cc** fields, for example.

- **Subject.** Here's the title of your message.

- **From.** Obviously, you know who this message is from—it's you. But you might have more than one email account; click here to choose which one you want for sending this message.

Now that you've set up the administrative details, you can compose the actual message. You can type, paste, or dictate the text. The toolbar (top right) offers some handy options:

- Aa reveals a formatting toolbar with options for font, size, color, style, and paragraph justification.

- ☺ opens a palette stocked with thousands of emoji symbols (page 141).

- 🖼 lets you choose a photo from your **Photos** collection. If you have an iPhone or iPad nearby, you can also use the **Take Photo**, **Scan Documents**, or **Add Sketch** commands described on page 276. It can be incredibly convenient to grab your phone and draw a little freehand sketch to add to your message—for example, to draw a map or diagram for somebody.

As you type, the Mac spellchecker marks questionable words with a dotted underline (see "TextEdit" on page 308).

If you run out of time or inspiration, you can choose **File→Save** to put your unfinished message into the Drafts folder. You can complete and send it whenever you have the time and interest.

But if you are ready to send the message, click ⌅ (or choose **Message→Send**). Mail sends the message and puts a copy in your **Sent** folder.

> **TIP:** If you open your **Sent** folder, click a message, and then choose **Message→Send Again**, Mail creates an outgoing message that's a duplicate of the one you already sent. You're free to edit this new message before you send it. For example, you might add a little nudge when you haven't had a response ("Just sending this again in case you missed it!"). This command is also handy when you just want to reuse the *shell* of a message—its subject, list of recipients, however you had the **Cc** and **Bcc** lines filled in, and so on—but you want to write an entirely different message.

File Attachments

Sometimes, the job isn't complete until you've attached a file to your email: some document or photo, for example.

To attach something to an outgoing message, click 📎 on the New Message toolbar, or choose **File→Attach Files**. The Open File sheet appears; from among the folders on your Mac, locate and double-click the file you want to include.

> **TIP:** You can choose more than one file to attach at the same time. Just ⌘-click or Shift-click them as though you were selecting them in a Finder window (page 112).

Or, if you can position your Mail window so you can see the desktop behind it, you can drag icons directly out of a window, or off the desktop, anywhere into the New Message window.

> **TIP:** Actually, you have a much better view of your desktop if you press ⌘-🖵—the Desktop Exposé feature—which makes all open windows jump to the edges of the screen. Now you can burrow through your folders to find the file you want to attach. Start dragging it, and then, without releasing the mouse, press ⌘-🖵 again to bring your message window back into view. Complete the drag into the message window.

The quickest way to attach a file is to drag it into the message from the Finder.

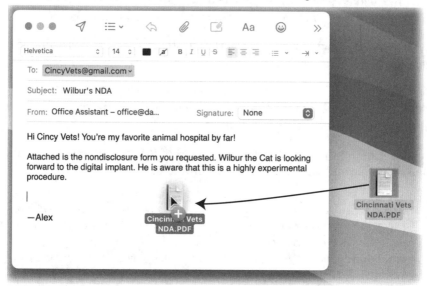

Dragging an attachment

You return to the New Message window. Your files' icons are clearly visible, to let you know they'll be attached when you send the email. (To remove an attachment, click its icon to highlight it, and then press the Delete key.)

> **TIP:** If you've added a graphic or a PDF document, you can mark it up with annotations before you send it. You might draw arrows on a map, for example, or circle something you want fixed on a rough draft.
>
> To reveal the markup tools—various pens and highlighters—click the attached image. Click the tiny ⌄ in its upper-right corner and choose **Markup**. (Or, if you have a laptop, force-click the attachment icon; see page 71.) Go nuts with your scribbles, and then click **Done**.

Signatures

You've seen signatures on people's emails, of course—the blob of sign-off text that gets stamped below their messages automatically. Usually it's their name and contact information, but sometimes it's a graphic or what the sender considers to be a hilarious quote.

To make up your own, choose **Mail→Preferences→Signatures**. You might be interested in having just one standard signature, but Apple has designed this screen so you can create many signatures and assign them independently to various email accounts.

Start by clicking **All Signatures** in the leftmost pane. To create a new signature, click **+**. Give it a name (*Standard work sig* or whatever) in the middle pane, and then edit its text in the rightmost pane. Use the Format menu to set up colors and formatting. Paste in an image, if you so desire.

Creating the signatures is only half the battle; now you have to make them available to your email accounts. To do that, drag a signature's name from the middle pane onto the *name* of the account in the leftmost pane.

But that's still only three-quarters of the battle. Now you have to specify which of the assigned signatures you want each email account to use automatically. In the left pane, click an account and choose from the **Choose**

HOW THE MAC SENDS HUGE ATTACHMENTS

The problem with file attachments is that most email systems impose a size limit for them: usually 5 MB. If you attach anything larger, your email will bounce back. It's a major pain—or at least it was until Apple invented Mail Drop.

This free feature—one of the iCloud services—lets you send gigantic files, up to 5 *gigabytes*, about a thousand times the usual limit. Now you can attach an entire video, a bunch of photos, or a large presentation without a care in the world.

The first time you attach a big file and then click **Send**, Mail asks: "Would you like to send this attachment using Mail Drop?" You would. Click **Use Mail Drop**.

Now, your recipient still has that 5 MB limit, so your Mac can't *really* send your

files as an attachment. What it does instead is upload your attachments to a temporary locker online, where they sit for up to 30 days.

If the recipients also use Mail, the app quietly downloads your attachment in the background, so it's on their Macs when they open the message. They may never realize what kind of shenanigans took place behind the scenes.

If you're sending to any other computer or mail app, though, the recipients see a **Click to download** link. When they click, their web browser opens, and the files begin to download.

Either way, Mail Drop is the biggest advance in email since the Delete key. ★

Each email account can have different signature options.

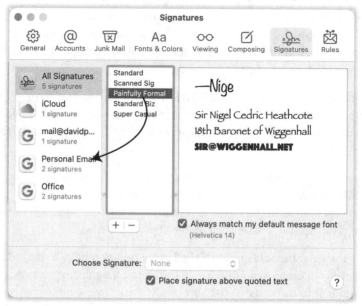

Email signatures

Signature pop-up menu. From now on, that's the signature you'll get every time you compose a new message in that account.

> **TIP:** If you choose **At Random**, Mail will paste a different signature each time you send a message. Live dangerously!

You can always override the automatic choice on an individual outgoing message; use the **Signature** pop-up menu.

Reading Email

As you become more popular, and your incoming messages arrive more copiously, a little numeric "badge" (**87**) appears, both on Mail's Dock icon and, in the Mail app, on the Inbox itself. In the main message list, blue dots draw your eye to the new messages.

Click a message once to view it in the message window, or double-click to open it into a separate window.

> **TIP:** Press ⌘-plus a couple of times to enlarge text that's too small, or ⌘-minus to make it smaller. (The same trick works in Safari.)

Conversation View

Unless you've changed the settings, your Inbox displays its messages in what Apple calls *conversation view,* and what the rest of the world calls *threading.* The idea here is that all the messages that are part of a single back-and-forth conversation appear as a single item in the main message list. Mail is smart enough to cluster messages in the same conversation, even if somebody changed the subject line along the way.

Expand or collapse the list of back-and-forths.

View the replies that led to this one.

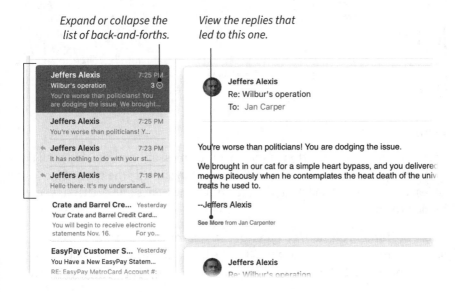

Expanding conversation view

When you click the most recent message of a thread, the message pane displays that message *and* all the earlier messages on an endless scrolling screen, as though they're printed on a roll of paper towels. To save you a lot of wading through clutter, Mail hides all the duplication and replies to replies.

> **TIP:** Should the most recent message appear at the top of the thread or at the bottom? That's up to you. Open **Mail→Preferences→Viewing** and turn **Show most recent message at the top** on or off.

In one regard, conversation view can be confusing. When it's turned on, Mail attaches all older emails on this topic to the most *recent* email. So if a particular conversation has been going on for six weeks, earlier messages no longer appear in their proper chronological place in the message list. Instead, they're grouped with the one that came in today.

If that drives you crazy, you can turn off **View→Organize by Conversation**. And sure enough: Now every message appears in its proper place, no longer grouped by conversation.

> **TIP:** Have you ever been caught in a **Reply All** maelstrom? Some idiot accidentally hits **Reply All** to the entire company, and then 275 people each hit **Reply All** to ask to be taken off the conversation, and pretty soon the entire network comes crashing to a halt because of geometrically increasing poor decisions?
>
> Yeah. It's bad. Especially if you've set up Mail to display a notification every time a message comes in.
>
> Fortunately, you can instruct Mail to stop notifying you about messages from a particular thread. Just select one of the messages, or a thread in the message list, and then choose **Message→Mute**.

Filters

When you click the ⊜ button at the top of the message list, Mail hides all the messages you've opened and seen. You're looking at only the unread messages. You've just turned on a *filter*; the other messages still exist, but you've hidden them.

While the filter is on, here's a handy trick: You can use the **View→Filters** submenu to reveal other ways to filter the list. (MacOS veterans note: You can no longer click **Unread** to change the filtering method.) You can view only

Click to turn filtering on and off.

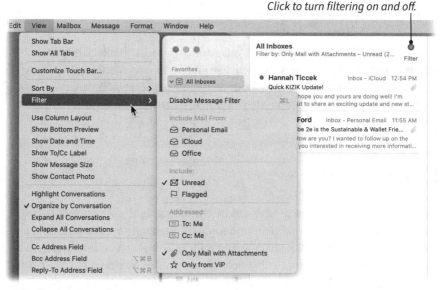

Filtering the message list

the flagged messages, only messages written directly to you (hiding the ones where you were only copied), only messages where you *were* copied, only messages with attachments, or only messages from people on your VIP list.

Click ⊜ again to turn off the filter.

> **TIP:** The **Message→Add Sender to Contacts** command creates a Contacts card for the person whose email is on the screen. The app even tries to fill in the phone number and address by seeing if that information is in the person's email signature.
>
> The advantage of putting somebody into Contacts: From now on, Mail can autocomplete their name when you're addressing an outgoing message and will no longer flag anything from this person as junk mail.

Opening File Attachments

Sometimes, people send you files attached to email messages, too. You will know because there's a little ⬦ icon next to the message's name in the message list, and also next to the sender's name when you open the message.

You also see an icon for the attachment at the bottom of the message. Here's a cool trick: Click it once and then press the space bar (or choose **File→Quick Look Attachments**). Like magic, a window pops open, showing what's inside that attachment—the full PDF file, photo, Word document, spreadsheet, or whatever it is. This is Quick Look, exactly as described on page 103—but now it's built into Mail, where you might find it even more useful.

If this document is valuable, you can free it from the clutches of the message that delivered it. Drag its icon out of the message window and onto any visible portion of your desktop behind it (or any visible folder or open window).

Or, if that seems a little imprecise, move your cursor to the top of the message without clicking. A ghostly set of buttons appears, one of them the ⬦ button. It's a pop-up menu whose commands include **Save All** (saves all the attachments to a folder you specify) as well as commands for each individual attachment, should you decide to preserve only one of them.

Data Detectors

As you work on your email in the coming months, keep an eye out for a dotted rectangle that appears when you point to an address, date, time, web address, or flight number. It indicates that macOS Big Sur has recognized what kind of information that is—and if you point to it and click the ∨, you get some special commands that know exactly what to do with that information.

For example, if Mail finds somebody's contact information in the message, that ⌄ button shows you a map; offers buttons that let you call, message, or FaceTime this person; and offers an **Add to Contacts** button.

And if Mail finds a discussion of an appointment, it opens an actual slice of your calendar, revealing the time slot of the appropriate day. The idea here is that you can see if you are, in fact, free then, and if so you can choose the appropriate calendar category (top right) and then click **Add to Calendar** to record the new appointment there.

Point to a phone number, address, flight number, web address, or time and date.

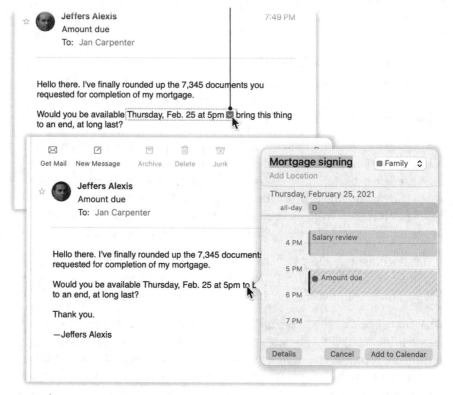

Data detectors

Surviving Email

Some people don't do anything at all with their incoming messages. Their Inbox counter grows to tens of thousands or even millions; they just ignore most of it. Lucky, lucky people.

The rest of us have to *do* something with our incoming messages. Here are your options.

MANAGE YOUR MAIL WITH MESSAGE RULES

Masters of the email universe don't drag messages into folders, type repetitive replies, and manually delete mail from obnoxious people. Instead, they teach Mail to handle these messages *automatically*.

That's the beauty of *message rules*: automated software robots that can file, reply to, or delete incoming messages according to certain criteria, like who they're from or what's in their subject line.

To create a message rule, choose **Mail→Preferences→Rules**. Click **Add Rule**.

In the resulting dialog box, use the options at the top to specify which messages to process.

For example, if you want Mail to handle all emails from a certain person, set up the first two pop-up menus to say "From" and "Contains," and then type the person's name or email address into the box.

Or, if you're sick of getting messages about property taxes, set the pop-up menus to say "Subject" and "Contains," and type *property taxes* into the box.

You're welcome to set up several such criteria, further refining the set of emails that Mail acts upon; click **+** each time you want to add another. (Use the pop-up menu to say either **any** or **all**, depending on how many of your criteria must be true for the rule to apply.)

Now, in the lower half of the box, specify what you want to *happen* to messages that meet the criteria. You can have them deleted, marked with a color, marked with a flag, moved into a folder, automatically forwarded, and so on.

Finally, in the box at the top, name your rule. Click **OK**. You've just created a mail rule. You programmer, you!

Mail now invites you to apply this new rule to all existing messages in the selected mailbox.

When you return to the main list of rules, you can drag your rules into the sequence you prefer. For example, maybe you want all messages from your mom to be moved into your Family folder first, *before* encountering the second rule, which auto-deletes new messages containing the word *recipe*. ✦

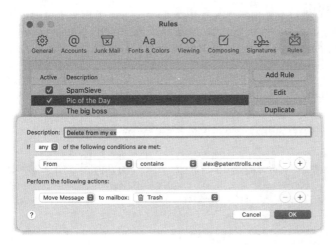

Replying

To answer a message, click **Reply** (↩) on the message toolbar (or choose **Message→Reply**).

Reply All (↩↩) is a little different: It sends your response to everybody originally addressed in the message, not just to the person who composed it. Reply All has gotten a lot of people in a lot of trouble over the years; be careful with it.

In either case, you get a new outgoing email message, already addressed, with the subject line filled in ("Re: [Whatever the original subject was]"). The original message—the one you're responding to—appears at the bottom of the window, as quoted text (in a special color and set off by a vertical bar).

You can, if you wish, use the Return key to create blank lines in that quoted message and splice your own responses into those gaps; see "Replying point by point." Your correspondent can see which parts are the original message and which are your response, thanks to the vertical line and colorized text of the original.

> **TIP:** Here's a fantastic trick: You can drag through a portion of a message—the part you want to talk about in your response—*before* you click Reply. When you do that, Mail pastes only *that portion* of the original message into the reply. Whoever reads your answer will know exactly what part of the original message you're talking about.

Now it's just like composing a fresh email message. Attach files, change the subject line, add or delete recipients, whatever. When you're finished writing your response, click **Send** (⌖).

Forwarding

Forwarding a message, of course, means sending one you've received along to a third person. To do that, click **Forward** (↪) on the toolbar (or choose **Message→Forward**).

A new outgoing message appears, ready for you to edit and address before sending. Usually you'll want to add a little comment at the top of it to explain why you're forwarding it. You know: "Dear Alex: See? I was right all along."

> **TIP:** If you choose **Message→Redirect** instead of **Forward**, you get a slightly different result: Your recipient sees the original writer's name as the sender, not yours. It's great for passing along an email that was sent to you by mistake.

Click where you'd like to insert a response, press Return to create a blank line, and type.

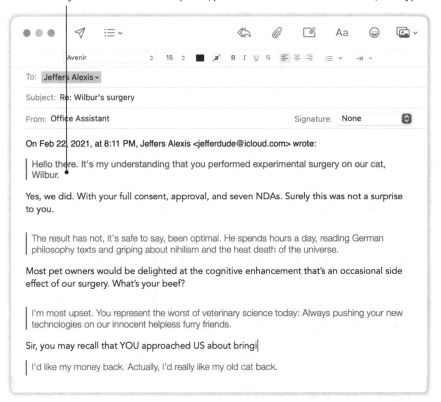

Replying point by point

Filing

The folders in the sidebar include such old favorites as **Inbox**, **Sent**, and **Drafts**. But you can create folders of your own, too, for organizing mail. You can create one called *Flight Confirmations*, one called *Stuff from Dad*, and another called *Deal with These Later*. You can even create folders *inside* folders.

To create a new folder, choose **Mailbox→New Mailbox**. Mail asks you to name the new mailbox. You can use the pop-up menu to specify which existing folder, in which email account, this one goes under.

When you click OK, a new icon appears in the mailboxes column, ready for use. You can file any message in the messages list by dragging onto that mailbox "folder."

Flagging

Mail offers seven colorful flags you can apply to messages to mean anything you want:

- **Make certain messages** stand out in the lists.

- **Round up deeply scattered messages** from your various folders and accounts into one place (the Flagged folder at the top of your folders list).

> **TIP:** You can expand the Flagged folder's flippy arrow to view subfolders named for your different flags: one containing messages you've tagged with the purple flag, one with the orange flag, and so on.
>
> This is also how you *rename* the flags, in case you don't find "Purple," "Red," and "Yellow" to be especially helpful. Just rename the subfolders of the Flagged folder.

- **Sort your mail list by flag color** (choose **Sort by→Flags** at the top of the messages list).

To apply a flag to a message, select it (or several) and then choose from the **Message→Flag** submenu. Alternatively, it's sometimes more efficient to

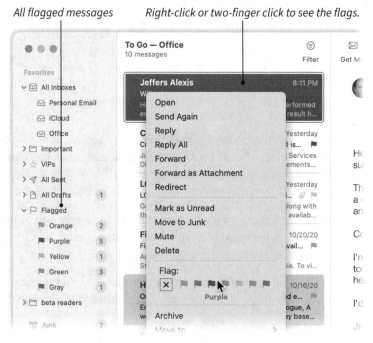

All flagged messages   Right-click or two-finger click to see the flags.

Flagging messages

right-click or two-finger click a selected message, or group of messages, and choose a flag from the shortcut menu.

(Use the **Message→Flag→Clear Flag** command to unflag a message.)

Searching

Mail isn't just an email-processing app; it's also, in a way, a database. Therefore, it's ideally suited to *finding* email in your ocean of thousands of sent and received messages.

To search all your mail, click the Q or the search box at upper right. Begin typing. Marvel as Mail hides all but the matching results, letter by letter.

> **TIP:** Do you want Mail to include messages in the Trash and Junk folders when it's searching? That's a choice you can make in **Mail→Preferences→General**.

In most regards, Mail's search feature works just like the one in the Finder (page 120). For example:

- **Mail understands plain-English search requests** like "mail from Sandy about merger" or "messages I sent last week."

- **Just below the search box,** Mail shows search refinements it thinks you might find helpful. For example, if you've typed *birthday*, Mail offers search shortcuts like **Subject contains: birthday**, **Mailboxes: birthday**, **Attachment name contains: birthday**, and so on.

  When you click one of those refinements, Mail places a little bubble in the search box—a *search token*, exactly like those described on page 128. You can create complex searches by creating several tokens in a row. Maybe the first one limits the search to messages with a certain word in the subject line; a second one might limit the search to a certain date.

> **TIP:** Once you've built up a complicated search using these tokens, you can save all that setup work for reuse later. Click **Save** just below the search box. Give the search a name and click **OK**. You've just created a smart mailbox, which appears in the list of mailboxes. Click that "folder" to see a list of messages that match the criteria at the moment.

You can search for certain text within an open message, too. That's what the **Edit→Find→Find** command is for.

Mail offers variations of your search term.

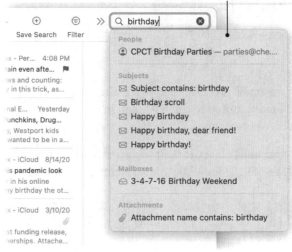

When you click a suggestion, a search token appears.

Mail search tokens

Deleting

To get rid of the message, click its name in the message list and then press the Delete key. Or, if the message is open, click **Delete** (🗑) on the toolbar.

> **TIP:** If you have a trackpad, you can also swipe leftward, with two fingers, all the way across a message in the message list to delete it. It feels a lot like the technique for deleting a message on the iPhone.

Mail's trash works like the Mac's: Things you put in there aren't really gone until you *empty* the trash. You have, in other words, a safety net, a little protection against mistakes. If you change your mind, click **Trash**; rescue a message by dragging it back into your Inbox (or any other folder).

To empty the trash, use **Mailboxes→Erase Deleted Items**. (If you have multiple accounts, choose **Erase Deleted Items→In All Accounts**.)

Actually, though, it's probably more convenient to set up Mail to empty the trash automatically on a schedule. Visit **Mail→Preferences→Accounts**, click the name of your email account, click **Mailbox Behaviors**, and adjust the **Erase deleted messages** pop-up menu. You might set it to empty the Trash every 30 days, for example, or every time you quit Mail.

> **NOTE:** Some email services, including Gmail, offer an **Archive** button instead of a **Delete** button. It's roughly the same thing, except that your deleted messages land in an **Archived** folder and don't auto-delete.

VIPs

In the worlds of high society, business, and politics, a VIP is a very important person. On the Mac, it's somebody whose *email* is very important.

That might be your boss, your spouse, your kids, or your state lottery notification department.

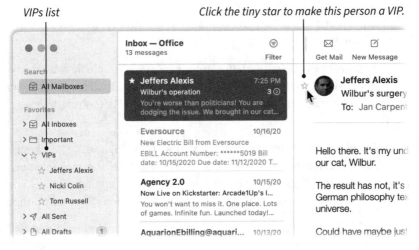

Designating a VIP

Once you've designated a VIP or two, a **VIPs** mailbox appears in your folder list. It always contains all the mail you've been sent—even deleted messages—from your VIPs. (Click the **>** to expand the list of VIP names, so you view their messages individually.)

> **TIP:** Thanks to iCloud syncing, your VIP setup appears automatically on all your other Macs, iPhones, and iPads.

If your VIPs are very, *very* important, you can direct the Mac to alert you with a notification when mail from them arrives. To set this up, open **Mail→Preferences→General→New message notifications**. Choose **VIPs**.

> **TIP:** And if you don't want any notifications to appear when mail comes in, open **System Preferences→Notifications→Mail**, and turn off **Allow Notifications**.

Dealing with Spam Mail

Spam is the worst. It's junk mail—marketing emails you didn't ask for.

Mail has a spam filter that's supposed to weed it out, but you have to train it. During your first few weeks with Mail, click any message you consider junk and then click **Junk** on the toolbar. Mail moves it to the Junk folder and gets smarter about that kind of message.

On the other hand, if Mail flags legitimate mail as spam (by coloring it brown), help it learn by clicking the message and then clicking **Not Junk**. Over time, Mail gets better and better at filtering.

(Mail never considers these messages junk: mail from people in your Contacts, mail from people you've emailed recently, and mail that addresses you by your full name. You can adjust these settings in **Mail→Preferences→ Junk Mail**.)

When Mail seems to have gotten smart enough about separating good mail from bad, turn on **Mail→Preferences→Junk Mail→Move it to the Junk mailbox**. From now on, Mail automatically puts what it considers spam messages into the **Junk** mailbox, where you can look them over and delete them all at once.

By the way: Did you ever wonder how the spammers get your email address in the first place?

Frankly, you gave it to them. Spammers use automated software robots that scour every public internet message and web page, recording email addresses they find. If you've ever provided your email address on any web page—signing up for a newsletter, buying something online, joining a discussion board—you've subjected it to spam.

The solution is to maintain two email accounts. Use one for all that public stuff. Use another for private correspondence—and never, ever type it onto a web page.

CHAPTER NINE

Safari

You may be too young to remember life before the web. You couldn't look anything up—you had to call places to get directions. You couldn't read articles on the web—you had to subscribe to newspapers and magazines made of paper. You couldn't order stuff online—you had to visit stores in person. Life was a living hell.

Safari is the Mac's built-in web browser. Its rivals, like Chrome and Firefox, have plenty of fans. But Safari is fast (even faster in macOS Big Sur), easy to use, and ingeniously integrated with the rest of macOS in ways that can save you time and hassle.

Five Ways to Start Out

The first time you open Safari, it has no idea what web page you want to view. You have to tell it, using one of these five methods.

Type an Address

That big box at the top of the screen is the address bar, and it's all-important. For starters, it's where you type the web address (the URL) of the page you want, like www.nytimes.com or www.whitehouse.gov.

The actual process goes like this:

1. **Click inside the box once (or press ⌘-L).**

   Whatever was already there is highlighted, ready to replace.

2. **Type or paste the address.**

   You can leave off the *http://* and *www*; Safari will add them for you. If you're trying to get to http:/www.amazon.com, for example, you can just type amazon.com.

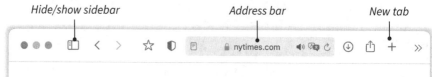

Hide/show sidebar · Address bar · New tab

Safari's layout

3. **Press Return.**

   In a matter of moments, the web page appears.

> **TIP:** When you're on a web page, the address bar shows you only the main part of
> its address. That is, you see only apple.com, not https://www.apple.com/macos/
> big-sur-preview/features/.
>
> If you don't find that shortening helpful, visit **Safari→Preferences→Advanced** and
> turn on **Show full website address**.

At this point, the Back button (<) lights up. It returns you to whatever
page you were just on before this. You can also press ⌘-◀ or, if you have

a trackpad, swipe left on it with two fingers. (On an Apple Mouse or older Mighty Mouse, use a *one*-finger swipe.)

And at *this* point, you can use the Forward button (>, or press ⌘-►, or two-finger swipe to the right) to revisit the page you were on when you clicked <.

You can also *hold down* the < or > buttons to see a pop-up menu of *all* the pages you've visited during this browsing session.

> **NOTE:** You can interrupt the downloading of a page by clicking ✕ on the toolbar (or pressing the Esc key).
>
> Once the page has loaded, that same button turns into a Refresh button (↻). Click it (or press ⌘-R) to re-download the page. That's useful when a page isn't looking or working quite right, or if you want to update it—a useful technique if it's a news or sports page that gets updated constantly.

Do a Search

Weirdly enough, Safari's address bar doubles as the search bar. It's built-in Google.

Just click in the address bar (or press ⌘-L), type what you're looking for (like *kidney stones* or *how old is the moon*), and press Return. Safari opens the Google results page.

A funny thing happens while you're typing, though. A menu of suggestions appears. It incorporates websites you've visited recently; websites you've

---

HOW TO SCROLL IN SAFARI

It's a rare web page indeed that fits entirely on your screen without scrolling. (*USCongressionalActsofBrilliance.com*, maybe?)

Therefore, scrolling is a critical web-browsing skill. And here's the first lesson: Dragging the tiny scroll bar with your cursor is the worst way to do it. It's fussy, it's imprecise, and it requires your hand on the mouse or trackpad.

Much better: Tap the space bar to scroll down one screenful. Press *Shift*-space to scroll back up. That trick works on every website, in every browser, and it gives you big, fat targets for your fingers.

Of course, the usual scrolling shortcuts also work. On the keyboard, you can press the PageDown and PageUp keys, if you have them. On a trackpad, you can drag upward or downward with two fingers. If you have a mouse, you can turn the scroll ball or scroll wheel—or, if it's a Magic Mouse, you can drag one finger up or down the top surface.

But really, use the space bar. Its invention was an act of brilliance. ✦

bookmarked; and suggestions from Wikipedia, Maps, and other sources. It's all in the name of trying to save you time and typing.

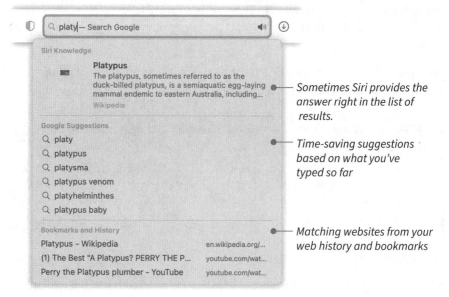

Sometimes Siri provides the answer right in the list of results.

Time-saving suggestions based on what you've typed so far

Matching websites from your web history and bookmarks

Safari search suggestions

Use the Start Page

The third way to begin your surfing session is to click something on the Start page, the page full of icons that appears when you open a Safari window without specifying an address.

You can also open the Start page whenever you want by clicking ⣿⣿ at the far left of the Favorites bar.

> **TIP:** Actually, what you see when you open a new blank page is up to you. Open **Safari→Preferences→General**; use the pop-up menus called **New windows open with** and **New tabs open with**. Your choices are **Start Page**, **Homepage** (whose address you specify by typing it into the Homepage box), **Empty Page**, or **Same Page** (meaning another copy of the page you're already reading).
>
> The choice is yours. Not everybody is on the **Same Page**.

The Start page gives you quick access to all kinds of websites Apple has concluded you care about. Here's what's here:

- **Favorites.** Here's a subset of your bookmarked sites.

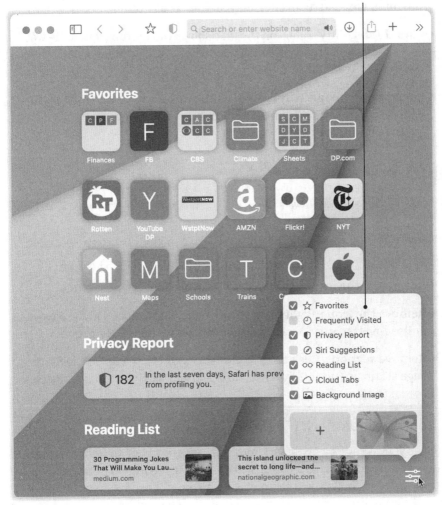

Turn Start-page categories on or off.

The Start page

- **Privacy Report.** Here Safari shows you how many websites it has prevented from tracking your activity online in the past seven days—a new macOS Big Sur feature described on page 215. Click here to see which ones they were.

- **Frequently Visited.** The latest slice of your History.

- **Siri Suggestions.** These are links to sites that macOS found in your Messages chats, recent Mail messages, and stories you've read in the News app. You know how you scroll through your messages or email, and people have sent you links, and you say to yourself, "OK, I'll click that later"? The Siri Suggestions section makes that easier.

- **Reading List.** These are websites you told Safari to memorize because you intend to read them later, maybe when you're offline. See page 206.

- **iCloud Tabs.** Nifty! Safari on the Mac shows you buttons for all the web pages that are currently open on your iPhone or iPad. iCloud synchronization strikes again.

- **Background Image.** Why not add a little whimsy to your life? If you turn on this option, the panel offers a selection of wallpapers (background images) for the Start page.

> **TIP:** Safari starts out showing you only some of the icons in each of these categories—for example, maybe a dozen of your favorites. To see the rest, point to them without clicking, and then click **Show More**.

If you do wind up relying on the Start page, you might want to customize it. For example:

- **Rearrange the icons.** You can drag them around.

- **Remove the icons.** Right-click or two-finger click an icon in the Favorites or Reading List areas; from the shortcut menu, choose **Remove Item**.

- **Remove sections.** Click the ⚏ at lower right to view a list of the Start-page sections. Turn off the checkboxes for any sections you don't need to see on the Start page.

Choose a Bookmark (Favorite)

Just as you can stick a bookmark into the pages of a book to find your place later, so can you bookmark pages of the web that you intend to revisit. In a browser's case, though, you can create as many bookmarks as you want— dozens or hundreds. You can make *folders* full of bookmarks.

In Safari, bookmarks are called *favorites*, but they work much the same as in other browsers. Here's what you can do:

- **Bookmark the page.** With the web page in front of you, choose Bookmarks→Add Bookmark (⌘-D). You get to name it, and you get to choose where it goes: into the **Bookmarks** menu, for example, or the **Favorites** bar described momentarily.

  Or use this more direct method: Point to the left edge of the address bar; use the ⊕ as a menu. Choose the location you want for your new bookmark. (With this method, you don't get the chance to rename the bookmark in the process.)

Where do you want this bookmark to appear?

Creating a bookmark

- **Organize bookmarks in the sidebar.** Click the ▭ button on the toolbar (or choose **Bookmarks→Show Bookmarks**). On the ▭ tab, you can view all your bookmarks—and organize them. For example, you can rearrange them by dragging them up or down the list.

  Some of the most useful commands are in the shortcut menu for a bookmark, which appears when you right-click or two-finger click it. They include **Rename**, **Edit Address**, and **Delete**.

  You can create folders (**Bookmarks→Add Bookmark Folder**) and then drag the bookmarks into them.

  (To delete a bookmark or a folder, right-click or two-finger click its name, and then choose **Delete**.)

- **Organize bookmarks in the editor.** If you have a lot of bookmarks, you may prefer to do your administrative work in the bookmarks editor window (**Bookmarks→Edit Bookmarks**). In here, it's even easier to delete one (click and then press Delete), rename one (just click it and type), or make a folder (**New Folder**).

- **Import bookmarks from another browser.** If you switched to Safari from Chrome or Firefox, Safari was supposed to offer to import your old bookmarks. If that didn't happen for some reason, you can always use **File→Import From→Google Chrome** or **File→Import From→Firefox**.)

- **Set up the Favorites bar.** Opening the Bookmarks menu and then clicking a bookmark—that's two clicks. Way too much effort.

  But if you install your favorite bookmarks and bookmark folders as buttons on the Favorites bar, you can get to them with only one click. (If you don't see the Favorites bar, choose **View→Show Favorites Bar.**)

  Menu method: Visit the web page you want. Choose **Bookmarks→Add Bookmark** (⌘-D). For the destination of the new bookmark, choose **Favorites.** Type a name for the bookmark, and then click **Add.**

Drag directly onto the Favorites bar.

Sites and folders on the Favorites bar

Mouse method: Drag the name of the current web page—in the address bar—directly onto the Favorites bar. (Don't click first—just drag directly downward on the address.) Once it's on the Favorites bar, type a new name for it (preferably a short one, since horizontal space is at a premium) and then press Return.

> **TIP:** You can jump to any of the first nine bookmarks on your Favorites bar by pressing Option-⌘-1 through Option-⌘-9.

You can rearrange these buttons by dragging them horizontally. Remove a button from the Favorites bar by dragging it down and away.

> **TIP:** In Bookmarks→Show Bookmarks, you can drag folders into the Favorites folder. They become pop-up menus of sites, as shown in the figure above.

Open the History Menu

Safari keeps track of every web page you've visited in the past week or so. It lists them in the **History** menu so you can revisit them later without having to remember how you got there. (It shows the most recent 18 or so individually; older ones it clumps into submenus by day of the week.)

CUSTOMIZING THE BUTTONS ON THE TOOLBAR

You've already encountered some of the standard buttons on the Safari toolbar—the ‹ and › buttons, the sidebar button (⬚), and so on. But considering how limited the space is on that toolbar, you should be pleased that it's easy to customize. Get rid of buttons you don't need. Add buttons you might find more useful.

Choose **View→Customize Toolbar**. A dialog box instructs you to drag any of the buttons in this catalog directly onto the toolbar to install them there. Which is great—if you know what they're for.

Here are some of the buttons whose purpose may not be obvious:

- **iCloud Tabs** opens whatever web pages you most recently had open on your iPhone or iPad, synced to your Mac via iCloud syncing.
- **Show/Hide Tab Overview.** If you become a fan of tabs (page 202), you can click this button to see miniatures of all open tabs, the better to find one you've lost in the sea of windows.
- **Home** opens whatever page you established as your home page (in **Safari→Preferences→General**).

- **Bookmarks** hides or shows the Favorites bar.
- **Zoom** gives you buttons that enlarge or shrink the web page before you.
- **Mail** opens a new outgoing message. The subject line contains the name of the site you were visiting; the body contains the actual web page.
- **Print.** Click to get a printout. (It's the equivalent of choosing **File→Print**.) As a courtesy to the planet, Safari automatically shrinks the printout by up to 10% if necessary to prevent printing an extra page containing only a couple of lines of text.
- **Privacy Report.** See page 215.
- **Web Inspector** opens a new window containing the HTML code for the page you're looking at.
- **Flexible Space** isn't really a button. It represents a gap you can install between *other* buttons on the toolbar, to separate clusters.

Incidentally: While the Customize Toolbar catalog is open, you can also drag any buttons *off* the toolbar to get rid of them. ✦

To revisit one of these sites, just choose its name. That much you probably figured out.

This part you never would've guessed: In the **Help** menu, there's a search box. When you type into it, Safari searches its own *menus*, including the History menu. In other words, by typing a word you remember from the title of a website you've recently visited, you can find a certain site buried among the thousands that Safari is remembering in the History menu!

Three Ways to Magnify a Page

It's not just you. The text on the modern web really is getting harder to read. Partly that's because today's screens have higher resolution than ever—more little pixels packed into less space—and partly it's because your eyes are getting older.

Starter text size   After a few presses of ⌘-plus

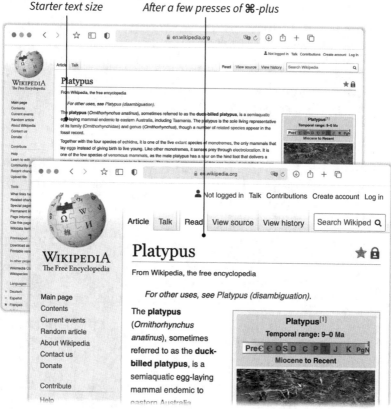

Enlarging the page

Fortunately, Safari offers all kinds of ways to magnify web pages:

- **Magnify everything.** Choose View→Zoom In (⌘-plus) or View→Zoom Out (⌘-minus) to make the entire page bigger or smaller. This is a fantastic trick that also works in Mail.

> **TIP:** If you have a trackpad, you can zoom in by spreading two fingers on its surface.

- **Magnify only the text.** If you press the Option key as you open the View menu, the commands you just read about change to say **Make Text Bigger** and **Make Text Smaller.** (Similarly, you can add the Option key to ⌘-plus and ⌘-minus.)

  These commands affect only the *text* on your page; graphics remain at their original size.

- **Set a minimum type size.** Open **Safari→Preferences→Advanced** and set **Never use font sizes smaller than** to, for example, 12 or 14 points. You've just prevented any web page from showing up with text that's too small to read. (Unfortunately, this setting doesn't affect web pages whose text size is hard-coded.)

Autofill

What we can do on the web these days is a miracle: Read the news, keep up with our friends, order groceries.

What's less of a miracle is how many of those activities require typing in your name, address, phone number, email address, and credit card number. Over and over, until your fingers are numb, calloused stumps.

Apple's answer to that tedium and repetition is Autofill. It's a feature that automatically completes those forms for you.

Contact Information

To set up Safari to fill in your name, address, and phone number, open **Safari→Preferences→AutoFill**. Turn on **Using info from my contacts.** Click **Edit.** From Contacts, find your own "card" to let Safari know where to find your information.

For the rest of your browsing career, a ⊙ button appears when you click into the contact-information boxes of the world's web pages. Click ⊙ for a menu of your addresses (like **Home** and **Work**). When you click, Safari fills in all the address boxes.

Or at least it's supposed to. It doesn't work on some oddball sites, and if it leaves a box empty here and there, press the Tab key to advance the cursor from box to box so you can type in the missing information.

Passwords

Safari can memorize your passwords for most websites, too, no matter how long and complicated they are.

It happens at the moment you first sign up for an account. You've just filled in your contact information and made up a password. You've clicked **OK**, **Next**, **Save**, **Continue**, or whatever the next button is. At this moment, Safari offers to save this new login information so you'll never have to type it manually.

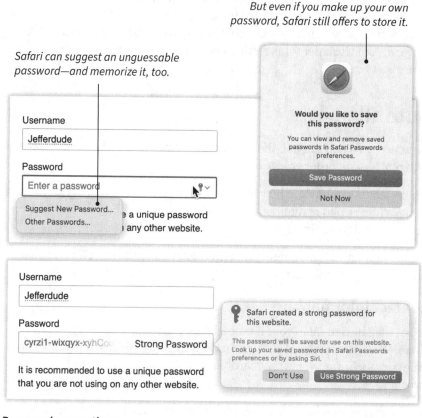

But even if you make up your own password, Safari still offers to store it.

Safari can suggest an unguessable password—and memorize it, too.

Password suggestions

SAFARI: THE PASSWORD CONSULTANT

Behind the scenes, Safari quietly compiles a master list of all the website passwords it's memorized. You can look them over, and even change them, at **Safari→Preferences→Passwords**. (You have to supply your Mac password or Touch ID fingerprint before you can see them. You know—for security purposes.)

But Safari does more than just memorize your passwords. It can actually guide you toward better password hygiene. For example, if you see a ⚠ next to a password, Safari is trying to warn you about a potential security risk. You'll see that symbol if a password contains a regular English word from the dictionary, which hacking software can easily guess.

Or maybe you've used the same password for more than one site. That's theoretically risky, because if the bad guys acquire *one* of your passwords in a data breach, suddenly they have access to *more than one* of your accounts. Since Safari can remember and fill in your passwords for you, what's the harm in making them all unique?

In macOS Big Sur, Apple takes the Safari password-guidance business one step further into Brilliance Land. When a big data breach does occur, Apple's servers know about it. They become aware of which passwords have fallen into nefarious hands.

Safari, believe it or not, runs a comparison of your saved passwords against the list of stolen ones. If it finds a match—if you are one of the unlucky victims whose passwords have been stolen—Safari lets you know while there's still time to change the compromised passwords.

Apple wishes to stress that all this computation is performed right on your Mac. At no time does Apple ever have any of your passwords. ✦

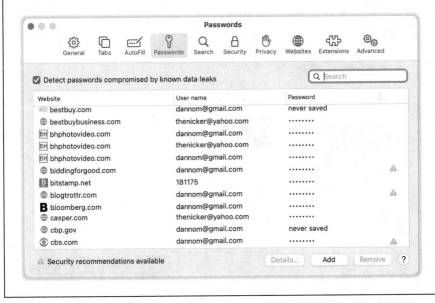

Actually, Safari even goes one better. If, when you're making up a password, you click the tiny 🔑 button at the right end of the box, Safari offers to make up a password *for* you. It's a ridiculously long, complicated, and unique one, impossible to memorize (or to guess). But since Safari is going to do the memorizing for you, what's the downside of accepting its unguessable suggestion?

You don't have to worry about what will happen when you try to access the site on your iPhone, iPad, or other Macs, either. They will *also* know this password and autofill it, thanks to the iCloud syncing feature in **System Preferences→iCloud→Keychain**.

Credit Cards

Safari can fill in your credit card information when you're shopping online, too—another time-saving blessing.

To store your credit card details, choose **Safari→Passwords→Autofill**. Click **Edit** next to **Credit cards**. Supply your Mac password or fingerprint.

And voilà: your list of stored cards. If you're just starting out, it's probably empty. Click **Add**. Now you have four little bits of data to fill in about your credit card: **Description** (*Citi Visa* or whatever), **Card number**, **Cardholder** (your name), and **MM/YY** (the expiration date). Click **Done**.

Click into the credit card box to see the menu of your saved cards.

Credit card autofill

One thing Safari does *not* ask for, and does not memorize, is the three- or four-digit security code, sometimes called the CVV or CVV2 code. You have to type that in every time you use this card on a web page. Apple says that extra step is a security thing, but you might consider it an annoyance thing.

The next time you're on a website that's requesting your credit card number, Safari sprouts a list of your saved cards. Click one to make Safari enter its stored information automatically.

Authentication Codes

If you hate nuisance and red tape, you're going to love this one.

These days, whenever you try to sign into an account where security really matters—your bank, your insurance company, your corporate network—supplying your name and password isn't enough. Most of these sites also use two-factor authentication (page 151): When you try to log in, the website

MacOS can auto-enter, into Safari, the code you receive in Messages.

Authentication autofill

sends a numeric code to your cellphone, which you're supposed to type in every time you sign in. You switch into Messages to view the number the website texted you, and you stare at those six digits, chanting them over and over, hoping you don't forget them by the time you return to your browser to type them in.

In Safari, there's no need. The instant that Messages receives the security code, a little box appears by your cursor, containing that code. Just click it to make Safari fill in the code. No memorization, no writing down, no panic.

Tabbed Browsing

Maybe you use only one web page at a time. Maybe you have several open in different windows. In those situations, you may never have an interest in *tabbed browsing*.

But millions of people use this feature to keep a bunch of websites open in a single, tidy window.

Tabs

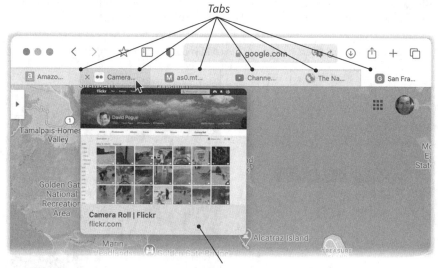

Tabbed browsing   *Tab preview (point without clicking)*

Safari, over the years, has become the mecca of tabbed browsing. Ready?

- **To create a new tab,** click the + at the top right of your window. You get what appears to be a new blank browser window, containing your home page, your Start page, or whatever you've specified in the **Safari→Preferences→General→New tabs open with** pop-up menu. Navigate away to whatever site you like.

Do notice, though, that you now have what appear to be two file-folder tabs at the top of the window—your original page and this new one. You can now click + again, and again, and again, creating more and more tabs in this single Safari window frame.

> **TIP:** You can also create a new tab by ⌘-clicking any link on any page. (If this doesn't seem to work, confirm that the first checkbox in **Safari→Preferences→Tabs** is turned on.)

- **Switch among tabs** by clicking the appropriate file-folder tab at the top of the window.

  Or, if you're a keyboard sort of person, press Control-Tab to advance from one tab to the next. (Add the Shift key to move to the *previous* tab.) You can even press ⌘-1 to open the first tab, ⌘-2 for the second one, and so on up to ⌘-9. Yes, each tab has its own private keyboard shortcut.

> **TIP:** You can drag your tabs horizontally to rearrange them. That's handy if you have an important one that isn't among the first nine that have keyboard shortcuts.

- **Preview a tab by pointing to it without clicking.** This feature shows you a miniature preview of what the tabbed content looks like—without actually leaving the tab you're on. (You can see this effect in "Tabbed browsing" on the facing page.)

> **TIP:** Does your laptop have a Touch Bar? Look down! When you're using tabbed browsing, it shows tiny tappable previews of all your open tabs.

- **Close a tab** by pointing to it without clicking, and then clicking the ✕ that appears at the left end. (If you close the *window,* you close all the tabs at once.)

> **TIP:** If you Option-click a tab's ✕ button, all the tabs close *except* the one you clicked.

- **Store a tab configuration as a bookmark.** Suppose you like to open the same set of tabs every morning: Gmail, a news site, Facebook, Twitter, and so on. It's possible to store all that as a single bookmark so they all load with a single click.

  Start by getting the window set up with the tabs you want. Now choose **Bookmarks→Add Bookmarks for These 7 Tabs** (or whatever the number is). Name your memorized set and store it wherever you like—in the Bookmarks menu or the Favorites bar, for example.

The next time you choose this bookmark, your brilliantly assembled tab configuration loads automatically.

- **See all your tabs at once.** On the toolbar, the ⬜ button opens the tab overview: a screen full of thumbnail images of all your tabs.

> **TIP:** If you have a trackpad, pinching two fingers inward also opens the tab overview.

The tab overview offers a few special features of its own. For example, if you've opened a dizzying number of tabs, you might wonder how you're supposed to find the one you're looking for. Well, here's an idea: With the tab overview open, simply start typing. As you type, Safari hides all the web pages whose names or addresses *don't* match what you've typed so far.

Recent sites open on your other Apple gadgets   Open tabs

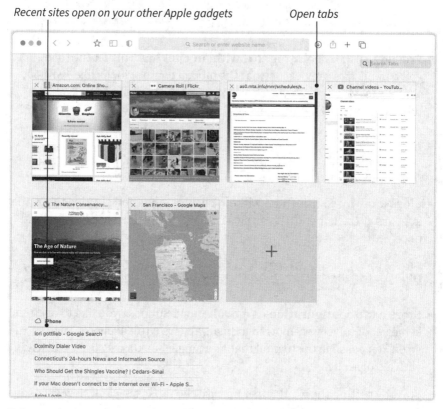

Tab overview

Furthermore, at the bottom of the screen, Safari shows you lists of all the windows and tabs you've had open on all your *other* Apple gadgets, like iPhones, iPads, or other Macs. Click to visit. (If this feature doesn't seem to be working, confirm that **System Preferences→Apple ID→iCloud→Safari** is turned on.)

- **Pin a tab.** If you drag a tab all the way to the left end of the tab bar—or if you choose **Window→Pin Tab**—it becomes a strange little square tab at the far left. Congratulations: You've just *pinned* the tab. At this point, it never goes away, even if you close the window or quit and then reopen Safari.

  Unlike a bookmark, a pinned tab is live. It's always updated in the background. When you click it, you see the latest incarnation of its web page. That makes it a natural for sites you revisit many times during the day, like Facebook, Twitter, or your email site.

NOTE: If you change your mind, use **Window→Unpin Tab**.

Drag all the way to the left.

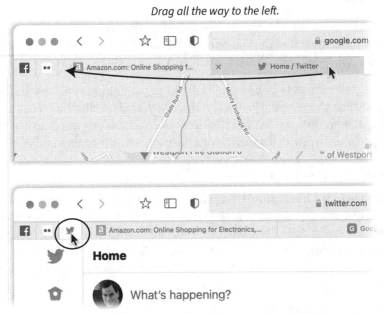

Pinning a tab

Ten Safari Specials

Apple is fully aware that people spend a lot of time in their web browsers, and that there are many web browsers to choose from. So over the years, its engineers have gone nuts fine-tuning Safari and adding grace notes to its feature list. Here are some of the best.

The Reading List

When you spot a web article worth reading but you don't have the time to finish it now, save it to your Reading List. The Mac downloads and memorizes the entire thing so you can call it up when you have more time—even when you don't have an internet connection, like on a plane or a deserted island.

To add a page to the Reading List, use one of these techniques:

- **Choose Bookmarks→Add to Reading List.**

- **Click + at the left end of the address bar.** It doesn't appear until you point to its spot without clicking.

- **Shift-click a link.** The linked page instantly appears in the Reading List.

Later, when you have time to do some reading, choose **View→Show Reading List Sidebar.**

The Reading List even keeps track of which stories you've read. If you scroll all the way to the top of the sidebar, you'll find **All** and **Unread** buttons that filter the list. (You'll also find a search bar.)

iCloud synchronizes your Reading List, too, to your other Apple machines. So the next time you're being jostled on the noisy subway, clutching your phone, staring at some article that you'd rather read at home on your Mac in peace, by a crackling fire, with a glass of wine, keep the Reading List in mind.

The Download Manager

The web is full of interesting things to download—not just software, but photos, PDF documents, and other goodies. But downloading is marred by two traditional pitfalls. First, what look like juicy downloads are sometimes malware—viruses, spyware, and other nasties. Second, downloads often land randomly into some computer folder, and you might not know where to look for them.

Safari addresses both of those problems.

First, whenever you try to download something from a new website, Safari asks if you're sure you want to permit this download. What it's thinking, but not saying, is: "You know, because people sometimes download bad things accidentally, or download from sites that aren't quite so savory, I just thought I'd make sure."

Do you want to allow downloads on "support.logi.com"?

You can change which websites can download files in Websites Preferences.

Cancel   Allow

Download warning

Second, you know very well where to find the stuff you've downloaded: Safari puts it all in your Downloads folder, which is listed in the sidebar of every Finder window.

But even if you can't remember that part, you've got the **Downloads** button (⤓) at the top of the Safari window. When you click it, a progress bar appears for each download that's still in progress, accompanied by a list of earlier downloads.

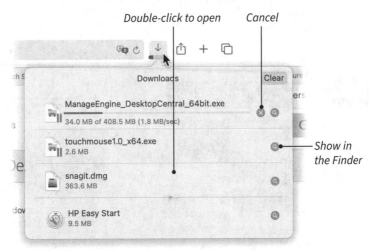

Double-click to open   *Cancel*

Downloads   Clear

ManageEngine_DesktopCentral_64bit.exe
34.0 MB of 408.5 MB (1.8 MB/sec)

touchmouse1.0_x64.exe
2.6 MB

Show in the Finder

snagit.dmg
363.6 MB

HP Easy Start
9.5 MB

Download manager

Downloading Internet Graphics

It's easy to save a picture you find on a website. Right-click or two-finger click it; from the shortcut menu, choose **Save Image to Downloads** or (to file it in some other folder) **Save Image As**. Or, if you prefer, choose **Add Image to Photos Library** to send it straight into your Photos app.

In each case, Safari grabs a copy of the photo from its place on that website.

> **NOTE:** The **Save Image** commands may not appear on websites that use nonstandard coding. No matter; you can always take a screenshot, as described on page 389.

Reader View

Truth to tell, websites have gotten pretty junky lately. They're filled with ads, banners, blinking things, self-playing videos, and other obnoxious detritus that distracts from what you're trying to read.

What you need is Reader view. When you click 📄 in the address bar, Safari hides *everything junky* from the web page you're trying to read. Only the text, headlines, and photos remain, against a simple white background and clean, clear type. (You can use the ᴀA button at the right end of the address bar to change the background color and the font.)

> **NOTE:** Often, Reader view bypasses the "Available only to subscribers" blockade screen on newspaper websites, but don't tell anyone.

Normal view (busy!)   *Reader view (ahhhh!)*

Reader view

To exit Reader, click 📄 again. Reader view is the best.

Saving Web Pages

Safari offers two great ways to save an entire web page to disk for long-term archiving.

First, you can choose **File→Save As**. In the Export As box, type a name for your archive, choose a folder for storing it, and specify the format. **Web Archive** is the option that preserves every part of the page—all its typography, layout, pictures, and videos.

The problem with saving a Safari page in this way is that only people with Safari can open it. If you send it to somebody who uses Microsoft Edge, for example, they'd have no idea what to do with it.

The second option, therefore, is to save the site as a PDF document (choose **File→Export as PDF**). This option, too, creates a perfect copy of the web page you're looking at—but as you know from page 296, a PDF document is universal. Anyone with any kind of phone, tablet, or computer can open it.

> **TIP:** And speaking of PDFs, here's a slick Safari exclusive that's worth remembering whenever you buy something online. Instead of printing your confirmation page, choose **File→Print** and, from the **PDF** pop-up button, choose **Save PDF to Web Receipts Folder**. Safari converts your receipt into a PDF file and saves it in a special folder called Web Receipts (in your **Home→Documents** folder), where you'll always know where to find it. It's the best.

Finding Text on Web Pages

You can search for words or phrases on a page, which is very handy if some article is long, cluttered, or badly designed. Just choose **Edit→Find→Find** (or press ⌘-F) to open the search box.

Safari highlights in yellow every occurrence of that phrase on the page and shows you how many matches it's found. Click the ‹ or › buttons to jump from one to the next.

Sharing a Page

It's easy enough to share a web page with someone. The 🔼 button at the top right of the window is a menu, and its options include **Email This Page**, **Messages**, and **AirDrop**—all good ways to send this page's address, as well as a thumbnail preview, to your pals.

The same menu offers options to send a link and preview into a **Notes** page or a **Reminders** item. These are great ways to memorialize important links.

PDF Pages Online

Safari is every bit as capable of displaying PDF files as it is standard web pages. You might not even be aware that you've opened one.

What you should be aware of is that once you *have* opened one, a PDF file is easy to save to your Mac. You might want to do that so you can look at it later in Preview (page 295), fill it out as a form, or make a printout of it.

Move the cursor around to make the download button and other controls appear.

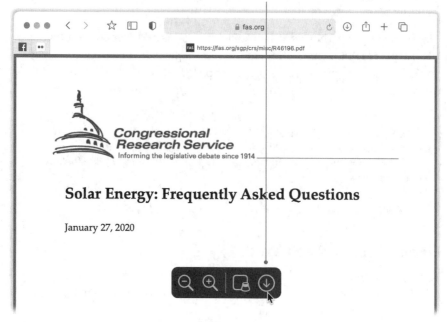

PDF files online

The trick is to move your cursor around near the bottom edge of the document. A special toolbar appears that lets you zoom out, zoom in, open the file in Preview, or save the PDF file to your Downloads folder with a single click.

Insta-Translation

The web is a global village, and not everybody's first language is English. Fortunately, in Big Sur, for the first time, Safari can translate an entire web page into your language with two quick clicks—as long as they start out in English, Spanish, Simplified Chinese, French, German, Russian, or Brazilian Portuguese.

Whenever you're on such a page, the 🗚 icon appears at the right end of the address bar. Click it and choose **Translate to English** (or to whatever language you prefer, assuming you've installed it using **Preferred Languages** in the same menu). Or choose **View→Translation**; same idea.

After a few seconds, the entire web page repaints itself in your language.

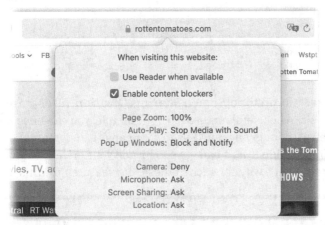
Settings for This Website

Safari can remember your settings for each website individually. That includes features like Page Zoom, meaning Safari can automatically enlarge websites that have tiny type. Safari can also prevent auto-playing videos—among the most annoying features on the internet.

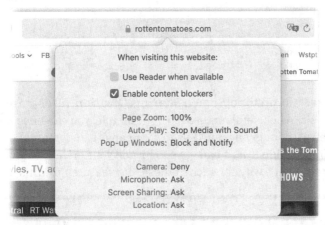

Settings for this website

To see the choices, choose **Safari→Settings For This Website**. Some of them are fantastic:

- **Use Reader when available.** Some websites are notorious for being so busy and cluttered that it's hard to find the material you wanted to read. This, of course, is why Reader view was born (page 208). This checkbox makes the website show up in Reader view every time you visit—automatically.

- **Enable content blockers.** What Apple means here is "ad blockers." These days, some web pages won't even load if you have an ad blocker turned on. Turning off this checkbox makes your ad blocker step aside.

- **Page Zoom.** Here's the option to set the enlargement (or enshrinkment) for a certain website automatically every time you visit.

- **Auto-Play** is for infuriating websites that begin playing some video the minute you visit. **Never Auto-Play** stops *all* videos from playing. **Allow All Auto-Play**, of course, blocks nothing. And **Stop Media with Sound** allows only silent or muted videos to play, which is less obnoxious.

- **Pop-up Windows** are wicked little windows, usually ads, that appear in front of your browser window. Or they appear behind it, waiting to jump out at you when you close your current window.

  This pop-up menu is your defense. You can **Block** those pop-ups, **Allow** them, or **Block and Notify** you that they've been blocked. That last option is useful on websites that might have a legitimate reason to open a pop-up window—like a ticket-selling site that wants to show you a map of the theater.

> **NOTE:** This feature blocks only pop-up windows that appear unbidden—not the ones that appear when you click a link, like an airplane seating chart.

- **Camera, Microphone, Screen Sharing, Location.** All of these Mac features can be useful on the web; for example, you can't use Zoom on the web without giving it access to your camera and microphone. But you certainly wouldn't want bad guys to invade your privacy through these channels. That's why, for each of these Mac features, for this website, you

---

A BROADER WORLD OF SAFARI EXTENSIONS

Extensions, in browser-ese, are plug-ins. They're little bits written by software companies that add new features to Safari.

The most popular extensions are ad blockers, which purport to keep ads off your web pages. (Many commercial sites simply refuse to load until you turn ad blockers off.)

But privacy tools, password-fillers, "save this page for later" features, and price finders are other popular options.

Finding and installing extensions has always been easy: In Safari, choose

Safari→Safari Extensions. You arrive at an App Store page dedicated to helping you read about, search for, and download extensions. Some are free; some cost a couple of bucks.

Once you download an extension, the final step is to open **Safari→ Preferences→Extensions** and turn on the new extension's checkbox.

What's new in Big Sur, by the way, is that Apple has thrown open the doors to software companies wider than ever; it's now easier for them to write Safari extensions. Soon, Apple promises, the online catalog will explode with juicy options. ★

---

can **Allow** the feature every time; always **Deny** its use; or have Safari **Ask** you for permission each time you visit.

There's one more useful component to this website Settings feature: In **Safari→Preferences→Websites**, Safari keeps a master list of all these features, along with which settings you've chosen for which websites. Taken together, this suite of options gives you a decent set of weapons against the worst of the internet sewer.

> **TIP:** Here's one more bonus Safari special: Safari can now play 4K YouTube videos (videos filmed in very high resolution). This marvel presumes, of course, that you have found a YouTube video that's *available* in 4K, and also that you have a screen capable of *displaying* 4K videos (Mac laptops, for example, don't have enough pixels). If all that checks out, click the ⚙ at the bottom of the YouTube video; from the shortcut menu, choose **Quality→2160p (4K)**.
>
> It looks pretty amazing; try not to drop your drink.

Privacy and Security

In the past few years, Apple has embraced privacy and security as its corporate mission on your behalf. And sure enough: Few web browsers offer as many features to protect you and your data as Safari.

You've already read about its password guidance (page 199) and its per-website settings for access to your camera, microphone, and location (previous section). But there's more.

Erasing Your History

Remember the History menu, which lists every website you visited in the past week? Sometimes, you may not want that list sitting there in full view of friends, family, or co-workers. You know who you are.

Fortunately, you can delete individual entries from the History menu. Choose **History→Show All History**, highlight the incriminating address, and then press the Delete key. Choose **History→Hide History** to exit the history editor.

You can also delete everything from the past hour, day, or two days. Choose **History→Clear History**, specify the interval, and click **Clear History**.

> **TIP:** In addition to the History menu entries, the Clear History command also vaporizes all the cookies (web page preference files) you've accumulated, your downloading history, and any cache files (web page pieces stored on your disk for quicker access the next time you visit).

Private Browsing

Every web browser these days offers private browsing. It's a mode that purports to maintain *no trace* of your web activity—no History entries, search histories, saved passwords, autofill entries, cookies, download history, cache files, and so on. You can surf the web in total anonymity, without leaving any tracks at all.

You can probably figure out why this feature is, ahem, so important to some people. Yes, that's right: so they can shop for their spouse's birthday present without risk of ruining the surprise.

In any case, all you have to do is choose **File→New Private Window** before you open the questionable site. (Any other windows can remain open, whether they're in private mode or not.) Safari reminds you that you're in private mode by turning the address bar dark gray.

When you're finished with your clandestine research, just close the window.

> **TIP:** You should definitely use private mode whenever you're using a public computer—in the library, for example—and visiting websites that require entering passwords or other information. You don't want the next guest to have access to your bank account, do you?

Cookie Control—and Privacy Control

A cookie is something like a preference file that a web page deposits on your computer so it will remember your information the next time you visit. Most cookies are helpful, because they spare you from having to reenter your name, your preferences, and other information every time you're there.

But if you're worried that they're somehow spying on you, you can always turn on **Safari→Preferences→Privacy→Block all cookies**.

Otherwise, Safari accepts cookies from sites you actually visit but rejects cookies put on your Mac by *other* sites—cookies from a fishy ad, for example.

The Privacy Report

In macOS Big Sur, Safari makes a radical effort to protect you from *trackers*: software scripts and cookies whose sole purpose is to harvest information about you and your activities. Most of this data is collected on behalf of advertisers for the purpose of targeting you with ads they think you'll be more likely to click. If you've ever done a search for, let's say, gym equipment and then discovered a bunch of ads for gym equipment in your Facebook feed later that day, well, that's trackers in action.

Apple can't stand trackers, and it wants you to hate them, too. What you do online should be your business, not the advertisers'. So in macOS Big Sur, for the first time, Safari automatically blocks trackers that "follow you across multiple websites and combine your activity into a profile for advertisers."

Safari does most of this invisibly. But if you'd ever like to see how hard it's been working to protect you from trackers, choose **Safari→Privacy Report**. (Or click **Privacy Report** on your Start page.)

Privacy report

Companies whose entire business is based on targeting you with ads tailored to your tastes—Facebook is a notable example—are furious with Apple for implementing this feature. Sit back and enjoy the show.

CHAPTER TEN

Messages and FaceTime

Text messaging turned out to be the killer app for smartphones—but texting actually makes a lot more sense on the Mac. Typing on a full-size keyboard is infinitely faster and more accurate than typing on a tiny piece of glass. Having a big screen makes it much simpler to read and juggle conversations. Having a computer attached to that screen gives you access to everything you might want to send in the text message—files, photos, videos, and so on.

And texting from your Mac means you retain all your text messages, from everybody, scrolling back in time forever. You can easily search them, copy them, paste them, forward them, and print them.

That's what the Messages app can do for you. Messages and FaceTime, Apple's video chat app, constitute the cornerstone of Apple-ecosystem communication—and they're the subjects of this chapter.

Welcome to Messages

Plenty of people don't even realize they can *do* text messaging from the Mac; they assume texting is a cellphone thing. But in fact the Messages app can send and receive both kinds of text messages:

- **Regular text messages.** You can exchange texts with any cellphone on earth. These show up in green bubbles in Messages. They're subject to all the usual limitations of text messages: For example, each one is limited to 160 characters, no choice of fonts or styles. But, hey—*any cellphone on earth.*

- **iMessages.** These are Apple's custom version of text messages; in the Messages app, these messages show up in blue. They're limited to exchanges between Apple devices; you can't send an iMessage to an Android phone.

On the other hand, that's about the *only* limitation. iMessages are like superhero text messages. They have no practical length limit—they can be pages and pages long. They can incorporate different fonts and text styles. You can send any kind of files—photos, movies, whatever you like, up to 100 MB in size—just by dragging them into the typing box. You can animate their delivery with special effects—confetti falling, balloons rising, and so on. (That's new in Big Sur; it used to be a phone-only feature.)

The iCloud service synchronizes your iMessages across your Apple machines, too. You can start texting on your phone while you're on the way home and then sit down at your Mac and pick up in midsentence. Your Messages chats look identical on every Apple gadget.

And because they're sent over the internet, rather than through the cellular network, iMessages don't count as text messages on your cellular plan.

More on the miracle of iMessages later in this chapter.

NOTE: iMessages are a feature of a free iCloud account. If you haven't already signed up for one of those, Messages will invite you to do so the first time you open it.

The Messages app itself has a few superpowers of its own. For example, it lets you transfer huge files to other people, neatly bypassing the usual file-size limits of email. It also offers a built-in screen-sharing feature (page 231) that lets you see and control the other person's Mac—an incredibly useful tool when you're trying to offer long-distance tech support.

How to Chat

The Messages screen has two parts. On the right, the current conversation, scrolling like a screenplay. On the left, a list of past conversations, each represented by the name and icon of the person you were chatting with and the most recent message you exchanged.

NOTE: In Big Sur, you can collapse the conversations list into a space-saving form that hides your correspondents' names and recent messages; only their icons are visible. To collapse the conversations list in this way (or to re-expand it), double-click the vertical divider line at its right edge. Or drag that line manually, if you're so inclined.

When you click a conversation at the left side, you see your entire history of texts with that person, scrolling all the way back to your first exchange.

NOTE: If it raises your paranoiac hackles to know that Messages remembers your entire message history with everyone, you can opt out. In **Messages→Preferences→General**, you can use the **Keep messages** pop-up menu to specify when old messages are automatically deleted. You can also delete an entire conversation by clicking it and then choosing **File→Delete Conversation**.

You can even delete one *individual* message by right-clicking or two-finger clicking it and, from the shortcut menu, choosing Delete. (Note that it disappears only from *your* copy of Messages. You can't delete one of your inflammatory texts from the *other* guy's copy. You've been warned.)

Pinned conversations   Conversations list   Attachments, settings, contact options

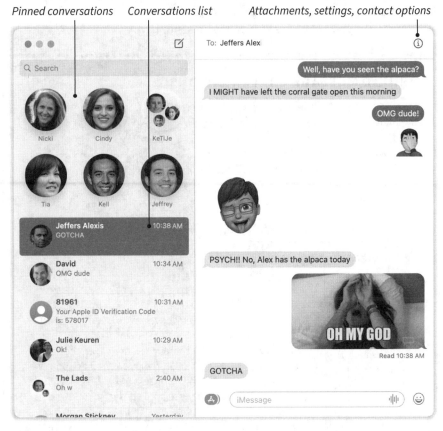

Messages layout

In other words, most of the time, you'll begin a chat session by clicking one of your *existing* conversations—and then just adding to it.

To message somebody for the very first time, on the other hand, click ☑ above the conversations list. In the **To** box, type the person's name or phone number. As you type, a list of potential matches appears. The names in blue are Apple people who can have an iMessages conversation with you; the ones in green are not.

> **NOTE:** Either an email address or phone number can be somebody's iCloud address. Most people have recorded both, so you can search for them by either bit of data.

If it's easier, click **+** to open a miniature version of your Contacts app, so you can find somebody's address using its tools.

Once you've chosen your correspondent, click in the box at the bottom of the window and begin typing. (The box starts out saying either **Text Message** or **iMessage**, to let you know whether this is an Apple person or not.) After each utterance you type, press the Return key to send it.

PINNING MESSAGE CONVERSATIONS

In macOS Big Sur, Messages includes a tiny change that makes a big difference. You can now *pin* certain conversation partners at the top of the conversations list—that is, install their icons there for good. In previous macOS versions, conversations could slide way down in the conversations list as more recent exchanges displaced them. Now you never have to hunt for the people you love. (Well, at least not in Messages.)

To pin a conversation, whether it's an individual or group chat, just drag it out of the conversation list into the Pinned area.

Or do it the long way: Right-click or two-finger click it. From the shortcut menu, choose **Pin**.

That person or group now enters a special VIP district for pinned people, above the conversations list. If you've signed into an iPhone or iPad with the same iCloud account, your pinned people appear there, too, courtesy of iCloud syncing.

Pinned conversation icons are especially informative. They display blue dots when they contain new messages, they display the blinking ellipsis (…) when someone's typing, Tapbacks (page 224) show up as they're applied, and so on. In group chats, people who've spoken up more recently get swelled heads. (That is, their faces appear larger.)

If one of these people or groups falls out of favor with you, just drag their face out of the Pinned area, or right-click or two-finger click and choose **Unpin**. ✦

When somebody wants to chat with *you*, the text appears as a notification (page 45). If you want to shoot over a quick reply, point to the notification bubble, click **Options→Show More**, and off you go.

If you'd rather open Messages for a full-blown conversation, click the notification bubble instead.

iMessage Specials

Whenever you're corresponding with another iCloud account holder, messaging becomes infinitely more flexible and powerful. As noted at the beginning of this chapter, there is no length limit to a message, you can use formatting, you can exchange files, and the whole thing is generally a lot more fun. Let us count the ways:

- **Informative feedback.** iMessages are inherently more informative than regular text messages. For example, if your correspondent is typing a message but hasn't yet sent it, an animated ellipsis (···) appears next to his speech bubble on your screen. It lets you know he's working on a reply and not just ignoring you.

> **NOTE:** The ··· dots remain on the screen for only 60 seconds and then disappear. If they go away, it could mean either that (a) the other guy has changed his mind about replying, or (b) he's writing a really long answer.

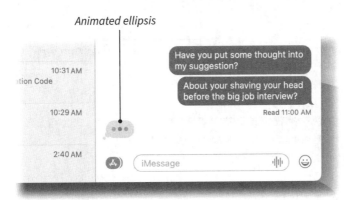

Animated ellipsis

Typing indicator

Similarly, Messages tells you when the internet has successfully delivered your message to the other person's phone or computer by displaying the word *Delivered* beneath it. That doesn't mean the other person has actually seen it, only that it has arrived on her machine. Either she or the machine might be asleep at the moment.

If you see the word *Read* beneath something you've sent, though, that means she has seen it, which can be an important piece of information.

What about you? Do you want the other guy to know when you've seen *his* texts? You can turn **Send read receipts** on or off in **Messages→Preferences→iMessage**. But you can also turn it on or off individually for each correspondent. You might want your love interests to know when you've seen their texts, for example, but you might not think your boss needs to know. The per-person override switch appears when you click ⓘ at the top-right corner.

- **Inline replies.** Ordinarily, texts appear in Messages in the order they're sent. That may seem logical, but it can lead to hilarious or disastrous

Right-click or two-finger click to view the shortcut menu.

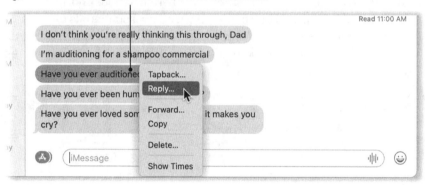

After you type the reply, it appears with the previous message it answers.

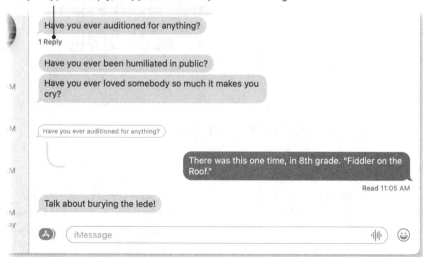

Inline replies

confusion if, because of typing lag, somebody's reply to topic A winds up appearing below topic B.

But in Big Sur, you can attach replies to one *particular* text message. The replies appear indented, and it becomes a subchat all its own.

> **NOTE:** If your conversation partners aren't using Big Sur or iOS 14, they don't see your reply as indented; it just appears at the bottom of the conversation as usual.

- **Live web and video links.** When somebody sends a web address in an iMessage, it shows up with a picture thumbnail of the actual web page. If it's a link to a YouTube video, you can play it right in Messages by clicking the thumbnail.

Click to play right here.   *Click to play on YouTube.*

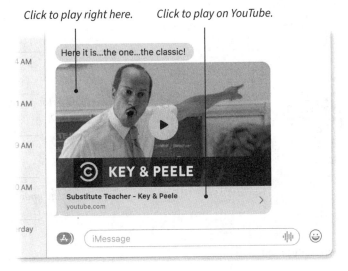

YouTube link

- **Adjust the text size.** In the **View** menu, a **Make Text Bigger** command awaits. (There's also a **Make Text Smaller** command, but who uses that?)

- **Where are you?** If your correspondent has turned on **Share my location** on her phone, then you can see a map of her present location by clicking ⓘ.

- **Searching.** Searching past chat conversations in Messages has been a sore spot for the Mac (also known as buggy and broken) for years. But no more: In Big Sur, Apple finally got around to making searching smart, fast, and useful.

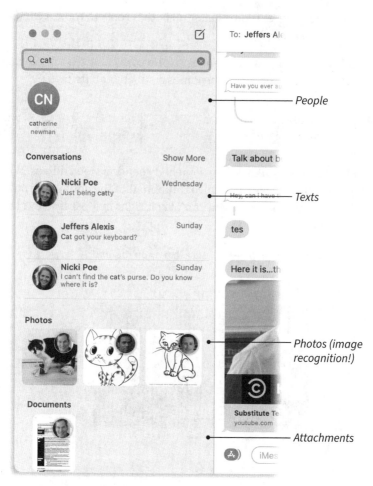

The new Messages search

Seven Ways To Be More Expressive

Mere mortal cellphones get to send typed text back and forth. But for you, the sky is the limit to your expressiveness:

- **Tapbacks.** In everyday texting, people spend an extraordinary amount of time sending standard reaction responses like "LOL," "haha," "Whaaaaaa??" and "!!!!!". But thanks to Tapbacks, you can respond to somebody's comment with one of six standardized reaction icons: a heart, thumbs-up and thumbs-down icons, a question mark, and so on. When you choose one, it appears instantly on your screen and your buddy's.

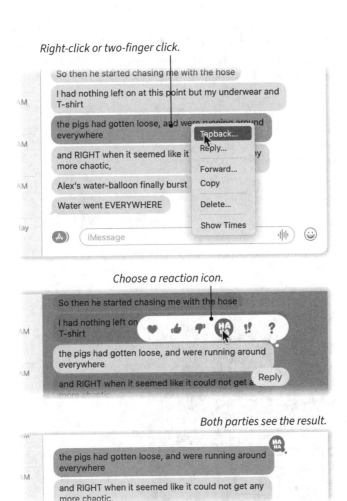

Right-click or two-finger click.

Choose a reaction icon.

Both parties see the result.

Tapbacks

- **#images.** It's become popular these days—and hilarious—to respond to something somebody has said with an *animated GIF*: a short, looping, silent video clip, usually a moment from a movie or TV show where the actor is exhibiting *exactly* the right reaction. On an iPhone or iPad, it's easy to attach these reaction GIFs—but Big Sur finally brings them to the Mac.

To look over your options, click the ⚘ icon next to the text box, and then choose **#images**. Type to search for the kind of GIF you'd like to send: *slow clap, popcorn eating, disgusted,* or whatever.

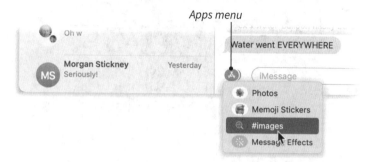

Apps menu

Type the reaction you're looking for; click an animated GIF to insert.

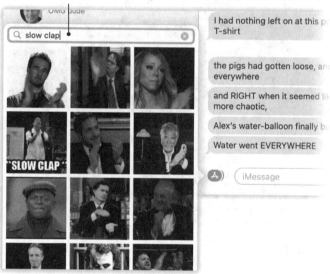

Finding a reaction GIF

- **Animated fanciness.** In Big Sur, you can finally send messages that show up on the other person's screen with full-screen animations: confetti falling, balloons rising, lasers shooting, fireworks blazing, and so on. (These sending effects, too, used to be limited to iPhones and iPads.)

To look over your options, click the ⚘ icon next to the text box, and then choose **Message Effects**.

Explore the horizontally scrolling row of animation styles; click one for a little preview of how it will look. You'll discover that **Slam, Loud,** and

Click an animated effect to preview it. Click to send.

Message effects

Gentle animate the typography of your text. **Invisible Ink** conceals your message with animated glitter until your recipient drags a finger or a cursor across it. It's a great way to reveal dramatic news, or to send very personal messages; you're granting your recipient the opportunity to choose a moment of privacy to reveal what you said.

The other options—**Love, Balloons, Confetti, Lasers, Celebration, Echo, Spotlight, Fireworks**—fill the entire background of the Messages window on the receiving end with animations. (In fact, if you send a text containing the words "Congrats," "Happy birthday," or "Happy New Year," Messages adds one of these animation effects *automatically*. No charge.)

NOTE: You can choose one of these effects even if you're not sending to a member of the Apple cult. But they won't see the animation; they'll just receive the possibly confusing phrase "sent with Slam effect," "sent with Confetti," or whatever.

- **Jumbo emoji.** No matter what kind of message you're sending, you can use the pop-up menu at the right end of the typing box to insert emoji symbols.

But in iMessages, if your message consists of one, two, or three emoji and no other text, you get to send them at three times their usual size.

Open the emoji picker.

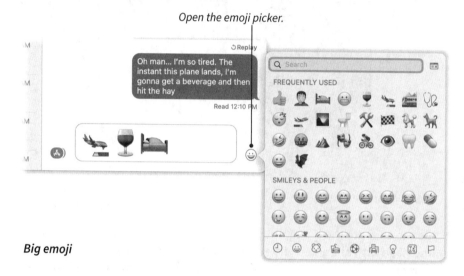

Big emoji

- **Audio texts.** Sometimes what you have to say is too long, too emotional, or too nuanced to send as text. Or maybe you can't type at the moment, because you need your eyes and one hand free—for example, when you're going up stairs. In those situations, you might welcome the fact that you can make a voice recording and send it right in the middle of your chat.

  Tap the ⦙⦙⦙ beside the text box and speak to record; tap ⦙⦙⦙ to finish recording. Now you can **Cancel** it or **Send** it.

- **Hop onto a call or video.** If you conclude that you would be better off switching to a phone call or a video call, tap ⓘ (top right of the window); on the panel, click 📞; click either **iPhone** (you'll be placing a standard phone call) or **FaceTime Audio** (which will be a voice-over-internet call that doesn't require an actual phone).

You can also hop onto a FaceTime video call by hitting the □◁ in that same ⓘ panel.

- **Make yourself a Memoji.** It may, at this point, be dawning on you that most of Big Sur's new features are actually old features Apple has brought over from iOS. Memoji fall squarely into that category. They're cartoon versions of yourself that you design—one facial part at a time.

  Once you've designed a reasonable facsimile of your face, Messages auto-creates dozens of variations, showing you in various poses and emotional states. You can use these—Memoji stickers, Apple calls them—as your responses to other people's texts.

  Start by clicking the 🅐 button, and then **Memoji Stickers.** The Memoji panel starts you off with various cutesy animal characters, but—unless your parents had a particularly unfortunate genetic mix—they don't look like *you.*

  Click the ● and choose **New Memoji.** The Memoji editor appears: a window whose sidebar lists various aspects of your appearance. Go through the screens—**Skin, Hairstyle, Brows, Eyes, Head, Nose, Mouth, Ears, Facial**

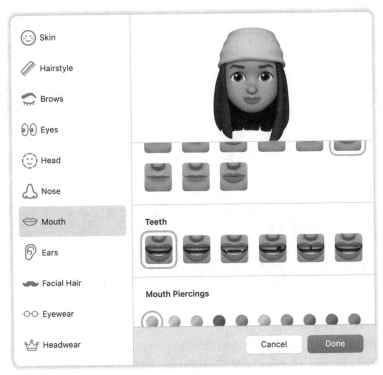

Memoji editor

Hair, **Eyewear**, and **Headwear**—and click the variation you think looks most like you.

> **NOTE:** That's what most people do, but there's no law that says your Memoji has to look like you.

By the end of the setup, you've created a little bubble-nosed, baby-headed version of yourself (or your idealized self). Click **Done**.

From now on, when you think the only perfect response to somebody's text is your Memoji head, click the ● button, click **Memoji Stickers**, and then click the variation you want. It pops into the texting box. You can type a comment, or you can just press Return to send it.

You're welcome to build other Memoji, too—as many as you want. They appear as horizontally scrolling miniatures at the top of the Memoji panel. To edit or delete one, click it, click ●, and then tap **Edit**, **Duplicate**, or **Delete**.

Group Chats

Messages is perfectly capable of accommodating conversations among more than two people. When you click the ☑, just enter several people's names or phone numbers in the **To** field, adding a space after each one. (Or, if you've set up a group in the Contacts app as described on page 256, you can type the group's name.) Group chats are highly handy for orchestrating get-togethers, collaborating on work or homework, or making any kind of plans.

> **TIP:** Sometimes it can be helpful to say something directly to one person in a group chat, instead of everyone. If you begin your text by typing the @ symbol, you get a pop-up list of participants; click the person you want. (Alternatively, just type the person's name.) The name turns bold and shimmers to indicate that Messages understands your meaning.

You can give your group a name—and, in Big Sur, a photo—which everybody in it sees at the top of the chat. Click ① and then click **Change Group Name and Photo**.

Now, there is a downside to group chats. If you've got notifications turned on for Messages, then you get an alert bubble every time a text message comes in. If you're part of a lively group chat, the notifications can drive you nutty, especially if most of it is chitter-chatter that doesn't concern you.

Choose a photo, emoji, or two-letter representation.

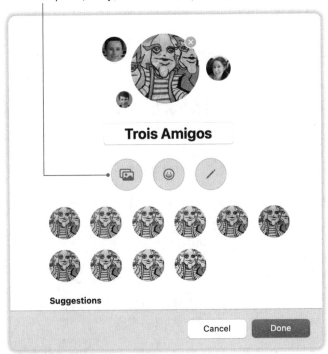

Changing a group photo

Fortunately, in **Messages→Preferences→General**, the **Notify me when my name is mentioned** option gives you a way out. It shushes all notifications from group chats except when somebody actually types your name.

Sharing Your Screen

It's tragic, really, that so few people know about the Messages screen-sharing feature. It's screen sharing across the internet. With permission, you can observe what's happening on somebody else's Mac, or even *operate* it, by taking control of its mouse and keyboard.

Screen sharing is an incredible troubleshooting tool. Next time someone you know is having trouble, you can parachute in virtually to see exactly what they're seeing, or to teach them what they're doing wrong. (The alternative—asking a nontechnical person on the phone to *describe* what they're seeing on the Mac screen—is enough to drive you both over the edge.)

Suppose, in this example, that you're in Boston, and you're trying to help your uncle in San Diego. Instruct him to open Messages; you do the same. Start a text chat.

Click ⓘ to open the details panel, click **share**, and then click **Ask to Share Screen**.

> **TIP:** If you choose **Invite to Share My Screen** instead, you're inviting your uncle to view and control *your* screen.

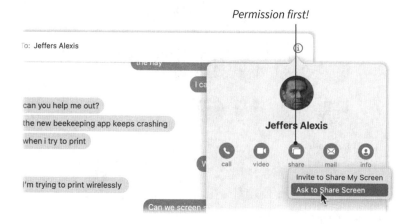

Permission first!

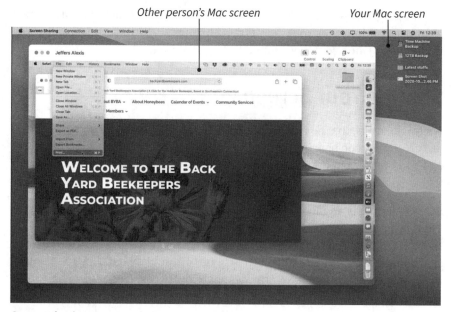

Other person's Mac screen

Your Mac screen

Screen sharing

On your uncle's Mac, a phone-ringy sound plays and a notification appears, letting him know you're trying to share his screen. He should click **Accept**. After a couple of seconds, the sharing begins. You see his screen, exactly as he sees it, in a window on *your* screen.

As a convenience, Messages also opens a two-way audio call. You and your uncle can talk to each other, so he can ask his questions and you can explain what's happening.

When the sharing window is frontmost, anything you do with your cursor and keyboard actually happens on *his* Mac. You can run his apps, copy and paste his information, surf his web, and so on.

Once you've solved his problem and accepted his expressions of gratitude, you can end the sharing session by choosing **End Sharing** from the ⬛ menu—or by choosing **Screen Sharing→Quit**.

Screen sharing: It's the next best thing to being there. And sometimes better.

FaceTime

FaceTime is Apple's video chat app. It's available on Macs, iPhones, and iPads, which makes it easy (and free) to make high-quality video and audio calls to anyone else in the Apple universe. It's perfect for letting parents or grandparents see the kids, for asking your spouse at home which kind of olive oil you're supposed to buy when you're at the store, and for long-distance relationships.

Making the Call

Often, a FaceTime call grows out of a text-messaging conversation. You know: "Will you bring my sweater when you come?" "I'm in the closet. Which sweater?" "The gray one." "There are three of them!" "OK, show me on FaceTime."

From Messages, click ⓘ to open the details panel, and then click **Video**.

The second-easiest way to start a FaceTime call is to use Siri (page 371). Just say, "FaceTime Dad" or "FaceTime Alex."

You can also start a FaceTime call from somebody's card in Contacts (click **FaceTime**). There's even a dedicated FaceTime app, which is almost identical to the Contacts app except that it lists *only* people who have iCloud accounts (and can therefore be FaceTimed).

On the call

Setting up a call

All (2) Missed

Enter a name, email, or number

+1 (203) 341-1200
Yesterday

+1 (203) 341-1200
Yesterday

EA Ernie Asprelli

FaceTime with Jeffers Alexis

Open "Add a person" panel
Mute
Hang up
Turn off camera
Go full-screen

Take a screenshot

FaceTime

No matter how you place the FaceTime call, what happens next is the same: The other guy's Mac, iPhone, or iPad rings, and a message indicates that you're trying to reach him for a FaceTime session. If he clicks **Accept**, the call begins. You see him on your screen; he sees you on his. (Each of you also appears on your own screen, in a tiny rectangle, so you can make sure you don't look ridiculous.) This is state-of-the-art videoconferencing.

During the call, you can make some adjustments:

- **Adjust the window size.** Drag any edge of the window, or go full screen by clicking the ⬤ button.

- **Move yourself.** Drag yourself—your little inset window—to another corner. That's handy if it's covering up something you want to see.

- **Turn the screen.** On iOS, the FaceTime picture is either portrait (upright) or landscape (widescreen), depending on how you're holding the phone or tablet. On the Mac—well, turning your laptop 90 degrees would be stupid. Still, you can choose **Video→Use Landscape**, or click the curved arrow that appears when you point to the inset that shows your face, to rotate the picture.

- **Mute yourself.** When you need to make an unseemly noise (or somebody in the background is making one for you), choose **Video→Mute**, or click 🎤 at the bottom of the video window.

- **Pause the video.** You can also mute the *video*, which is handy when you need to tend to something the other person doesn't really need to see. To do that, click 📹; your image goes black on the other guy's screen. He can still hear you, though.

> **NOTE:** The Mac also pauses the call when you minimize the FaceTime app or send it to the back. (That's for one-on-one calls only, not group calls.)

You can unpause the call by bringing the FaceTime window back to your screen.

Click ✕ to end the call.

Group FaceTime

FaceTime isn't exactly Zoom. Nobody is going to use it to make presentations to 100 people at a time.

Still, up to 32 people can join a FaceTime call, as long as they're using macOS or iOS versions from 2018 or later.

Here again, one natural way to begin such a session is from a group chat in Messages. Click ⓘ at the top of the screen, then **Video**; you've just sent each member of the group an invitation. Anyone who clicks **Join** becomes part of the call.

You can also use the FaceTime app to begin a group call. In the **To** box, enter the names of the participants, and then click **Audio** or **Video**. Their Macs, phones, or tablets start ringing. Once they click **Accept**, the call begins.

Each person's face appears in its own micro-window on your screen, which enlarges itself when they're speaking. (If there are more than four participants, the fifth and additional people appear in a scrolling row at the bottom of the screen.)

If you click someone's box, it gets a little bigger and identifies their name or iCloud ID. If you *double*-click someone's box, it fills the window so you can really have a good look.

Once the call is underway, you can add another participant by clicking the screen, clicking ···, and then clicking **Add Person**.

FaceTime Audio

Even today, 10 years after the dawn of the video-call era, most people still make phone calls. Audio, no video. Not for technological reasons, but for human ones. Video calls burden you with looking good and seeming put together; audio is often less of a hassle.

FaceTime audio calls work just like FaceTime video calls. To start in Messages, click ⓘ, and then **call**, and then **FaceTime**. In Contacts, click **call**, and then **FaceTime**. In the FaceTime app, click 📞 and then **FaceTime Audio**.

You'll discover that FaceTime audio calls sound amazing. The audio has a clarity and presence that sounds more like a CD than a lousy cellphone call.

> **NOTE:** When someone calls *you*, a notification bubble appears, and you hear a ringtone. (It's whatever sound you choose in the FaceTime app, in **FaceTime→ Preferences→Settings**.) You can even answer the call on the Lock screen if your Mac is asleep, without having to enter your password. You "pick up" by clicking **Accept**.
>
> If you click **Decline**, the caller sees a "not available" message for you. And if you never want people to bug you with FaceTime calls, open the FaceTime app and choose **FaceTime→Turn FaceTime Off**.

CHAPTER ELEVEN

Apple's Five Multimedia Stores

I n 2007, Apple Computer dropped the "Computer" from its name. The point was obvious: Its intention was to become a phone company, a tablet company, a services company, a subscription company. Today, in fact, Apple makes more than twice as much money from selling digital goods (music, movies, apps, subscription services) than it does from *Macs*.

No wonder, then, that your Mac comes with five apps dedicated to helping you find (and buy) digital multimedia goods: Books, Music, News, Podcasts, and TV. They're named and designed after the corresponding iPhone and iPad apps; in fact, most of the Mac's multimedia-store apps are "ports" (code adaptations) of exactly those apps.

Luckily for you, all five of these sibling apps use the same design and layout: a navigation sidebar on the left; a search box at the top; and, in the main window, thumbnail images representing the goodies on offer. Each store remembers what you've bought; you can feel free to delete stuff (books, movies, apps) from your Mac, confident that you can re-download them at any time. (When you see a ☁ on a thumbnail, it means "You own this; just click to download it again.")

Books

You've probably heard of Amazon's Kindle books, the ebooks that have taken over the world. Well, Apple wasn't about to sit on the sidelines. It has created its own version of an ebook store—and an ebook reader app. This is it.

On the first tab, called **Books,** you see the thumbnails (miniature images) of all the books you've bought (if any).

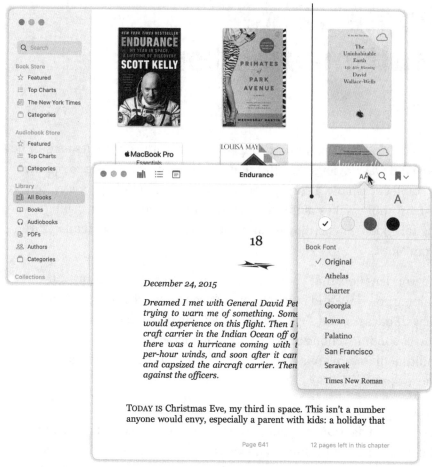

Type size, font, and background page color

The Books app

Double-click a book to open it on the screen and begin reading.

> **NOTE:** Thanks to the miracle of iCloud syncing, the Apple ecosystem remembers where you were in each book, no matter which Apple gadget you pick up to read it.

If no books appear here, then your first job is to shop the bookstore.

Try Before You Buy

The first few tabs (**Featured, Top Charts,** and so on) are the Apple Store itself. You can use the search box (and then click **Store**) to seek by name or author, and you can click one of the miniature covers to read reviews and a description. Also on the description page: the all-important **Get Sample** button, which downloads one chapter for free, so you can get a sense of whether you'd even like the book.

If you love it, click the price button to buy and download it.

The **Audiobook Store** tabs work exactly the same way, but for audiobooks. Listening to books read aloud by professional narrators is a great time-passer when you're working out, cleaning the kitchen, or driving.

> **TIP:** When you're listening to an audiobook in Books, don't miss the ••• button. It offers a playback-speed control—great when the narrator is reading too slowly for your brain's superior processing power. There's also a **Sleep Timer**, which stops playback after a certain amount of time, so you can fall asleep to a book being read to you.

Reading a Book

Once you've opened the book, you can turn pages either by pressing the space bar—or, on a laptop, by swiping two fingers to the right across the trackpad.

> **TIP:** On a laptop, the Touch Bar (page 73) shows a map of the entire book. Drag your finger across the Touch Bar to scroll through the pages. But don't worry that you've now lost your place; a handy ↻ button on the Touch Bar teleports you right back to where you began.

The Books app is teeming with supplementary features (to view the toolbar that contains them, move your cursor to the top of the window):

- **Change the typeface or the page color** by clicking the AA button.

- **Bookmark a page** by clicking the ▮. Later, you can return to any page you've bookmarked by using the ▮ as a menu.

- **Search the book** using the Q icon.

- **Highlight some text** by dragging across it and then clicking one of the color dots. This is also your chance to **Add Note**—a floating sticky note.

> **TIP:** All your highlighting and notes sync to all your *other* Apple gadgets—your iPhone, iPad, and so on.
>
> Fortunately, they don't appear in anybody *else's* copies of this book. That would get really annoying.

- **Listen to the book read aloud** by choosing **Edit→Speech→Start Speaking**. It's not *exactly* an audiobook, but it's pretty close, and it's free; Apple's synthesized voice is fairly realistic and even includes fake breaths. (**Edit→Speech→Stop Speaking** makes it stop.)

Music

The Music app, a simplified version of the jukebox app once called iTunes, has two purposes in life. First, it's the dashboard for Apple Music, Apple's version of Spotify. You pay $10 a month for the privilege of listening to just about any music ever recorded, streaming over the internet. Second, you can use Music to manage and play your own music files—MP3 files and whatnot that reside on your Mac.

Apple Music

The cool thing about being a member of the Apple Music service is that you can instantly listen to any album, band, or song among the 60 million in the Apple Music catalog, or you can ask Siri to "Play the top songs of 2020" or "Play some good jogging music" or "Play Billy Joel." Or you can listen to ready-made playlists in every conceivable category.

The less cool thing is having to pay $10 a month forever.

In any case, the first three tabs in the sidebar are primarily intended for subscribers. **Listen Now** and **Browse** are scrolling billboards of performers, albums, and songs Apple recommends for you; **Radio** is a list of simulated radio stations, each of which perpetually plays a certain style or performer.

Your Music Collection

Even if you're not an Apple Music subscriber, the Music app can still be useful—as a jukebox app to manage any music files you own, like MP3 files you've imported ("ripped") from your CD collection.

You can import these files from your Mac's drive into Music by choosing **File→Import**. Or, if they're still on CDs, insert the disc into your external CD

For Apple Music subscribers only

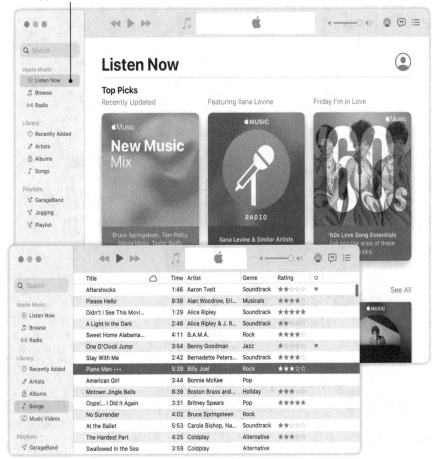

The Music app

drive, if you have one. (A drive like this costs about $25.) The Music app invites you to begin importing the songs on it.

Once you've got some songs on your Mac, you now know the purpose of the Library categories: **Recently Added, Artists, Albums, Songs,** and **Music Videos.** Here's where you can slice and dice your music collection.

Playlists

You can also use the Music app to organize your *playlists*—memorized lists of songs that play sequentially. You might have one for a dinner party, one for jogging, and so on. What's interesting is that you can make your playlists from either Apple Music streaming songs or music files you actually own.

Start by choosing **File→New→Playlist**. At the bottom of the sidebar, a playlist appears, ingeniously named **Playlist**. In the main window, rename it something descriptive.

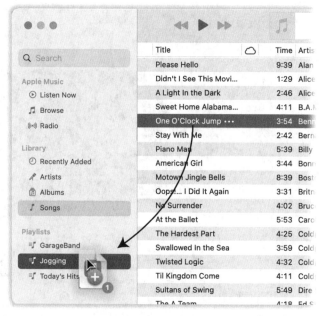

Making a playlist

Now click one of the Library headings—**Artists** or **Songs**, for example—to find the songs you want to add to this playlist. Drag their names onto the name of the new playlist.

To play the playlist, click its name and then click ▶ at the top of the screen. Or just ask Siri to "Play my Makeout playlist" (or whatever it's called).

News

This app rounds up news headlines from hundreds of online newspapers, magazines, and websites. You can browse sections like Entertainment, Politics, and Science. You can search for a topic. Or you can choose **File→Discover Channels & Topics** to *tell* the app which publications and topics you care about, in effect constructing your own custom newspaper.

Over time, News is supposed to tailor the stories it's suggesting according to your own tastes, by studying which articles you actually wind up clicking to read.

Podcasts

This app may look like Apple's other digital-media app stores. But this time, what you're "shopping" for is podcasts—those free "radio shows," distributed exclusively online, produced by everybody from professional radio production companies to amateurs in their living rooms.

Podcasts, as you may have heard, are spectacular audio companions when you're working out, commuting, cleaning the house, cooking, or doing anything else that requires your eyes but not the full capacity of your brain (or ears).

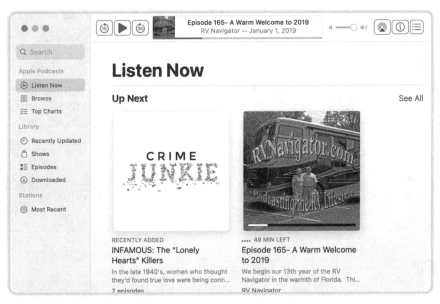

The Podcasts app

If you point to the "cover" of a podcast, a ▶ button appears so you can listen to the latest episode with a single click. If you click the cover instead, you open up the details page for this podcast. Here you'll find a description, reader reviews, a list of episodes—and, maybe most useful of all, a **Subscribe** button.

Subscribing

Most podcasts are series; there's a new episode every week or so. Subscribing tells your Mac to download every new episode automatically.

Once you've subscribed to a podcast or two, visit the Library section of the sidebar. **Shows** lists the individual podcast titles, **Episodes** is a list of their latest shows, and **Downloaded** reveals which episodes are already on your Mac. That's an important consideration if you're about to spend a long, boring time where there's no internet—like on a transatlantic flight.

> **TIP:** On the details screen for any podcast for which you have a subscription, you can click ••• and then **Settings** to reveal a vast number of controls. You can specify how many episodes you want downloaded, when you want them deleted, how often you want the Mac to check for new episodes, and so on.

Handily enough, whatever podcast setup you create on your Mac (subscribing, downloading, or listening) is magically mirrored on your iPhone or iPad, so you can pick up right where you left off.

Playing Podcasts

As you might guess, clicking the ▶ button on any podcast or episode thumbnail image begins playback. There's a volume slider at the top of the screen, but maybe even more important there are ⑮ and ㉚ buttons next to the ▶ button. They let you skip backward 15 seconds or ahead 30 seconds at a time, which is ideal for zooming past ad breaks.

> **TIP:** In **Podcasts→Preferences→Playback**, you can change the skip duration for these two buttons. It can be 10, 15, 30, 45, or 60 seconds per click.

TV

This strange little app is the front end for Apple's online movie and TV-episode stores.

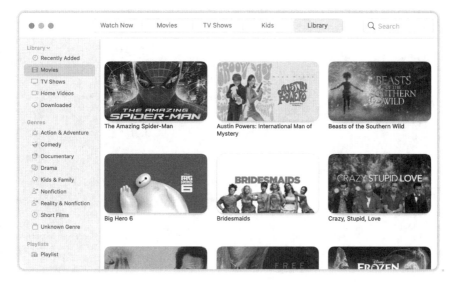

The TV app

The tabs across the top boil down like this:

- **Watch Now** lists a bunch of suggestions for things to watch from Apple's paid video service and others, including Showtime, CBS All Access, and Starz.

- **Movies, TV Shows, and Kids** let you buy individual TV episodes (usually $2 each) and rent or buy movies (usually $10 or $20 to buy, $3 to $5 to rent). Once you rent a movie, you must start watching it within 30 days, and then you must finish it within 48 hours; after that, it disappears from the app.

- **Library** is where you'll find all the TV episodes and movies you've bought or rented from Apple. Not all of them are ready to play at this moment; if you point to one of the thumbnails and see a ☁ icon, it means you have the right to watch it, but the actual video is still online. You'll have to download it, at least partially, before you can start playing it.

> **TIP:** Keep in mind that you can view whatever you're watching on your TV, which may be much bigger and nicer than your computer screen; see page 28.

CHAPTER TWELVE

The Other Free Apps

W hen you bought your Mac, Apple gifted you with a cornucopia of starter software. In your Applications folder at this moment are over 50 apps that let you get all kinds of work done: movie and music playback, photo management, word processing, a spreadsheet, Calendar, Notes, Reminders, and so on.

The big internet apps (Mail, Safari, Messages, and FaceTime) and Apple's multimedia-store apps (Books, Music, News, Podcasts, and TV) get their own chapters in this book; this chapter offers mini-manuals for the best of the rest.

> **NOTE:** Each of the major Mac apps greets you, upon its first opening in Big Sur, with a welcoming "splash screen" that advertises some of its new features. To avoid boring you silly, this chapter doesn't mention them. Just click **Get Started** or **Continue** to dismiss that welcome screen and open the app.

Calculator

For straightforward calculations, you don't need this app. You can just ask Siri things like, "What's 15% of 24?" or "What's 1,723 times 7?" You can also do math in the Spotlight search box (page 120).

Still, the Calculator app has a few tricks up its sleeve. For one thing, it's three calculators in one. Using the **View** menu, you can make it the standard four-function **Basic** calculator, a **Scientific** calculator, or a **Programmer** calculator that lets you work in hexadecimal.

Using the **Convert** menu, you can ask the Calculator to convert units for you: temperatures, currency, grams, inches, meters, miles an hour, whatever.

Calendar

Calendar fulfills the usual function of a paper calendar but exploits its electronic nature by entering recurring events automatically, popping up reminders when the time comes, and allowing you to share certain sets of appointments with other people electronically.

The Four Views

Using the tabs above the calendar, you can switch among **Day**, **Week**, **Month**, and **Year** views. Each view scrolls endlessly into the future or the past, either horizontally or vertically. (Use the left and right arrows at the top right of the screen, scroll with two fingers on your trackpad, or scroll using your mouse.)

As you might guess, the larger the time period you see on the screen, the less detail you can see for each day. By the time you're in **Year** view, all you see are individual date numerals, although you're welcome to click one to see what's happening that day.

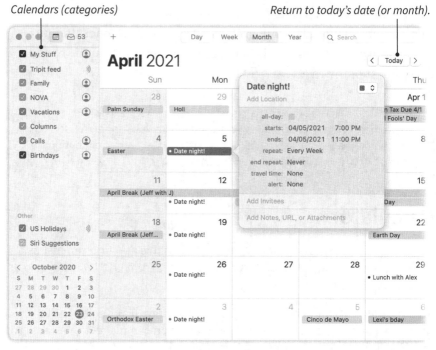

The Calendar app

Subscribing to an Account

It's quite likely that, as a human being living in the 21st century, you already have an online calendar, courtesy of Google, Yahoo, Facebook, or Apple's own iCloud. The Calendar app on your Mac is perfectly suited to showing those calendars to you.

All you have to do is choose **Calendar→Accounts**. Click the + below the list. The Mac shows a list of services whose calendars it can import: **iCloud, Microsoft Exchange, Google, Yahoo,** or **AOL**. (There's also an **Other CalDav Account** option, for use with oddball services not listed here.)

Click the service name, click **Continue**, enter the email address and password you use to log into that service, ensure that **Calendars** is turned on, and click **Done**.

At this point, the list of calendar categories at the left side of the window shows the name of the service you've just added. (If you don't see this sidebar, click 🗓 at the top left of the window.) And all the appointments on that online calendar now appear in the Calendar app on your Mac.

And you didn't have to manually enter a single appointment!

Recording an Appointment

When you do want to enter an event on the calendar manually, the exact steps depend on which view you're using:

- **Month view.** Double-click any blank spot on the appropriate calendar square and type the name of the event. Inside the New Event bubble, you can also change the times for this appointment, specify its calendar category (described in a moment), and edit other details.

- **Day view, Week view.** For a one-hour appointment, just double-click the starting time slot. If the appointment isn't exactly an hour long, drag vertically through the time slots to specify when it starts and ends. In either case, the New Event bubble appears, ready to accept the details.

- **Any view.** In Year view—or any of the other views, for that matter—you can click + at top left to open the Create Quick Event bubble. Here you can type, in plain English, the name and time of your appointment: *birthday party March 9*, for example, or *report deadline Weds 8 PM*, or *date with Chris 7:30-8 PM*.

    When you then press Return, the app interprets and parses what you've typed.

What you type *What Calendar proposes*

Creating a Quick Event

If you don't specify any time, Calendar assumes this is an *all-day* event, which will appear as a banner on the calendar. And if you enter a time but not a date, the app assumes you mean today.

> **TIP:** If you *hold* your cursor down on the **+** instead of clicking it, you get a menu of your calendar categories (page 253), so you can specify which category this appointment belongs to. If you just click the **+** instead, the new appointment winds up in your *default* category (**Calendar→Preferences→General→Default Calendar**).

The instant you press Return, Calendar displays the new, parsed event on the appropriate calendar date, with its details bubble open. Here's where you can specify further details, as described next.

> **TIP:** Actually, the fastest way to enter an appointment is by voice. Use Siri (page 371) to say, "Make an appointment called *Lunch with Alex*, Tuesday at 1 PM." Siri shows you the appointment she's about to record, and you just click **Confirm**. Boom: It's done.

The Details Bubble

All the methods of creating a new appointment on the calendar conclude with the appearance of the details bubble. Depending on your level of fastidiousness, you can take some time to record the finer points of your new appointment:

- **Specify the calendar (category).** You can turn entire categories of appointments on or off—showing or hiding them all on the calendar—with a click. That's a big help in reducing calendar clutter. See page 253 for details.

  In any case, this pop-up menu lists all the categories you've created, so you can choose one for your new appointment.

- **Add location.** As you begin typing an address, Calendar searches its internal atlas and offers to complete the address for you. If you choose one of its suggestions, Calendar shows you a map of the place and the current weather.

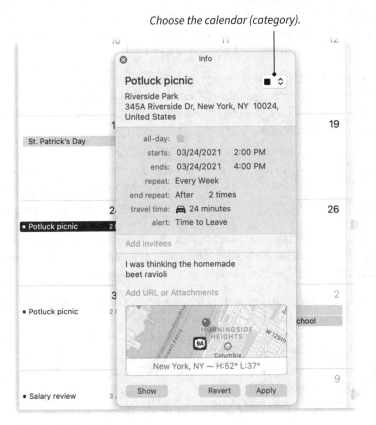

Choose the calendar (category).

The details bubble

If you click **Add Alert, Repeat,** or **Travel Time**, the bubble expands enough to reveal some new controls:

- **All-day** means this appointment has no particular time of day associated with it. It's a holiday or birthday, for example.

- **Starts, ends.** If this *isn't* an all-day event, you can use the **starts** and **ends** controls to dial up the exact time span. You can, of course, click the segments of the time and date readout, but it's often faster to press the Tab key to jump from block to block, and then just type the digits you want.

- **Repeat.** If you click here, you can turn this appointment into one that reappears on the calendar automatically—**Every Day, Every Week, Every**

Month, or **Every Year**. Or you can click **Custom** to input some wacky non-standard repetition schedule.

You can also tell Calendar when you want it to stop repeating this appointment—either **After** a certain number of times, or **On Date** that you specify. (You can also choose **Never**, meaning this appointment will show up on your calendar until the end of time. That would be a good option for, say, a tax deadline or anniversary.)

- **Travel time.** If Calendar recognized the address you entered, it does something rather ingenious: It shows the amount of time it will take to get there. You get to see the time involved for all the available travel options. (Note to procrastinators: It's *travel time*, not *time travel*.)

> **NOTE:** This automatic calculation doesn't work unless the Mac knows where you're starting from. It uses your home address or work address, depending on the time of day, as you've recorded it on your own "Me" card in Contacts (page 256).

If you *haven't* entered an address for this appointment, or you overrode the Mac's autocomplete suggestion, you can choose one of the travel times in the menu, like **15 minutes**, **2 hours**, or whatever.

- **Alert.** Your Mac can get your attention when you have an appointment coming up. Use this pop-up menu to specify how much warning you want. If you choose **Custom**, you can specify not only wackier time intervals, but also what else you want to happen when the time comes: You get sent an email, for example, or a certain file opens automatically. Using this feature, you could set up Calendar to throw up on the screen your written remarks for the meeting that begins in 10 minutes.

- **Add invitees.** In the world of business meetings, this feature is bread and butter. In this box, type the email addresses of everybody who's invited to the meeting; press Return after each one. (Or, if they're already in your Contacts, just type their names.) When you click **Send**, everybody gets an invitation to the meeting. If they click **Accept**, two things happen: The appointment gets entered onto their *own* calendars, and green check-marks appear next to their names on this panel, so you can see who all will be there.

- **Add Notes, Add URL, Add Attachments.** This is your chance to store any random bits of information (or even a file from your Mac) that will help you remember what this appointment is about.

Once you've recorded all these details about the new appointment, if you have any energy left, click anywhere else on the screen to close the bubble.

Editing Appointments

To change an appointment's details, double-click it to reopen the details bubble.

To reschedule an appointment, it's much more fun to just drag the appointment to another date or time. You can do that in Day or Week view.

> **NOTE:** If you're editing a repeating event, Calendar asks if you intend to change only *this* occurrence, or this and all *future* occurrences.

In Day or Week view, you can also make a scheduled meeting longer or shorter by dragging the bottom edge of its block.

And to delete something completely, click it once and then press Delete.

Calendar Categories

Apple chose perhaps the most confusing possible term for the color-coded categories you can use to organize your events: *calendars*. Ugh.

You don't have to use this feature; it's fine to record all your appointments in a single category. On the other hand, you might want to use them to separate your calendar appointments into categories like Work, Family, Trips, and so on. The advantage is that you can hide or show all the appointments in

CALENDAR CATEGORIES THAT NOBODY ELSE CAN SEE

When privacy is an issue, there's one other useful aspect of calendar categories: You can store all the appointments in a given category either in *iCloud* (on Apple's online servers)—or *On My Mac* (never online or anywhere else).

Storing a calendar in iCloud is useful because it means the Calendar app on any Apple gadgets you use—other Macs, iPhones, iPads—show the same set of appointments and categories.

But On My Mac is a better choice if you're planning a surprise party for someone,

you don't trust the hackers of the internet, or you're having an affair.

To indicate where you want a calendar stored, right-click or two-finger click it in the Calendars list; from the shortcut menu, choose **Get Info**, and then click **Calendar Info**. On the **Alerts** tab, use the **Account** pop-up menu to specify where you want this calendar stored.

At this point, your secret will be revealed only if somebody happens to look over your shoulder. ★

a category just by turning off its checkbox in the list at the left side of the window. (If you don't see this list at the moment, click the little ▦ icon near the top left of the window.)

To create a new calendar, choose **File→New Calendar**. Now there's a new item called *Untitled* at the bottom of the Calendars list. Type a name you prefer and then press Return. (If you ever want to rename a calendar, just double-click it.)

You can change the color of any existing calendar by right-clicking or two-finger clicking and, in the details bubble, choosing one of the color dots. (You can also click **Custom Color** to choose some offbeat color of your own.)

Name the new category and then press Return.

Right-click or two-finger click to change the color.

Click ⊙ to share the category with other people.

Setting and sharing a calendar category

Publishing Calendars Online

Here's one of the payoffs of using the calendar categories feature: You can post a certain calendar online so other people can see it or subscribe to it, which adds its appointments to their own calendars. It's perfect for posting the schedules for a club, sports team, or group of colleagues.

To share a calendar, proceed like this:

1. **Point to a calendar's name without clicking; click ⊚.**

   The sharing-details bubble appears.

2. **Click Share With, and type the email addresses of anybody who should be able to see this category.**

   To enter several people's addresses, press Return after each address and keep typing.

> **TIP:** If you turn on **Public Calendar**, you'll create a special web version of this calendar, viewable (but not editable) by anyone who has the web address shown in the box. Click the 🖰 button to send that URL to anybody you like, by mail or text message.

3. **For each participant, specify whether or not they're allowed to make changes.**

   Click the tiny ⌄ next to an email address to see your options: **View Only** (this person can see the appointments but can't change them) or **View & Edit** (they can change or delete things just as you can). Use these powers wisely.

4. **Click Done.**

Behind the scenes, your Mac sends email invitations to the people you have blessed with access. If someone accepts the invitation, then your shared calendar shows up on their computers, and on your Mac, a green checkmark appears in the sharing panel next to their names.

> **NOTE:** From time to time, *you* might be the recipient of an invitation sent by somebody *else*. In that case, the little ✉ icon at top left sprouts a number, letting you know somebody has invited you. Click it, and then click **OK** in the invitation to add that calendar category to your own calendar.

Chess

The Mac comes with only one game, but it's a classic. When you open the app, you're set up to play against the computer. You go first; just drag the piece of your choice to begin the game.

It's 3D chess! Drag the board to rotate it in space.

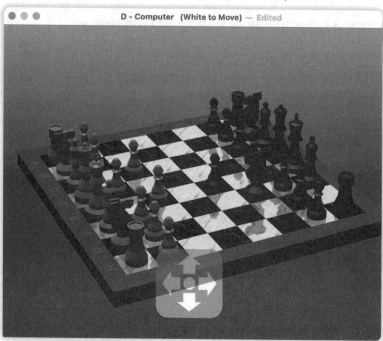

Chess app

If you choose **Game→New**, you can specify other variations, like **Human vs. Human** and **Game Center Match** (meaning you want to play against one of your friends online).

> **TIP:** Chess first appears as though the board and pieces are carved from handsome wood. But in **Chess→Preferences**, you have a choice of simulated materials: **Wood**, **Glass**, **Marble**, or (this is not a joke) **Grass**.

Contacts

Here's the Mac's master address book—the central people database that your *other* apps consult for information. This app, too, can sync its data, via iCloud, with your iPhone, iPad, Apple Watch, other Macs, and even Windows PCs.

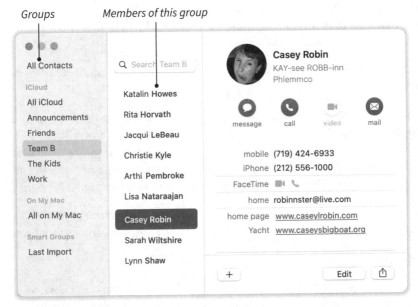

All Contacts

iCloud
All iCloud
Announcements
Friends
Team B
The Kids
Work

On My Mac
All on My Mac

Smart Groups
Last Import

Q Search Team B

Katalin Howes
Rita Horvath
Jacqui LeBeau
Christie Kyle
Arthi Pembroke
Lisa Nataraajan
Casey Robin
Sarah Wiltshire
Lynn Shaw

Casey Robin
KAY-see ROBB-inn
Phlemmco

message   call   video   mail

mobile (719) 424-6933
iPhone (212) 556-1000
FaceTime
home robinnster@live.com
home page www.caseylrobin.com
Yacht www.caseysbigboat.org

+   Edit

Contacts

Importing Addresses

If you've got a stash of addresses in a service like Google, Yahoo, Microsoft Outlook, or Microsoft Exchange, you'll be able to save a lot of time.

In Contacts, choose **Contacts→Accounts**. Click the name of the service, log in, and turn on the Contacts checkbox. From now on, your Contacts app incorporates that online address book automatically.

Entering Addresses

Like a paper Rolodex, the individual "pages" of contact information are called *cards* in Contacts.

To start fresh for a new person, choose **File→New Card**. Type in this person's details; press Tab to hop from field to field. As you'll soon discover, there's room for a lot more than name, company, phone numbers, email addresses, and mailing addresses. Contacts offers space for web addresses, birthdays, notes, and even the ringtones you want to hear when this person calls or messages you.

You can have multiples of certain fields, too, because lots of people have more than one email address or phone number. That's why, as soon as you fill in *one* phone number, another **Phone** field appears, empty, just beneath. Use the blue pop-up menus next to each field to indicate which *kind* of number it is (**mobile, home, work**, and so on).

Actually, there are even more oddball fields available. What if you want to record this person's Twitter handle, department, birth name, or the phonetic pronunciation of their name?

That's all available, too. Choose the new field you want from the **Card→Add Field** menu.

Finally, don't miss the **Picture** tab, where you can click the **+** to choose one of the **Defaults** (symbolic photos like animals and flowers), **Camera** (take a photo right now), silly shots you've taken with **Photo Booth,** or anything in **Photos**. Choose one and hit **Next**; now you can reposition the photo or (with the slider) adjust its size within the circle. Click **Save** when the headshot looks right.

Once you've assigned a photo to a card, you'll see it when the person calls, emails, or texts, both on the Mac and on any other Apple gadgets you own.

Editing a Card

When you first open Contacts, every card is locked, so you don't change or delete any precious information by accident. Find the card you want to

---

ORGANIZING YOUR CONTACTS WITH GROUPS

In the far-left sidebar, Contacts lists whatever address books you've imported from Google, Yahoo, and so on. But it also lists whatever *groups* you've created.

A group is a bunch of related cards, saved under a single name, like Work Friends, The Kids, or Poker Pals. Next time you're addressing an email in Mail, for example, you can type just that short name; Mail fills in everyone's email addresses automatically, even if it's dozens of people.

To create a group, first make sure you've selected the right account to put it in (Google, Yahoo, iCloud, whatever)—if you have more than one.

Choose **File→New Group**. Type a name for the group in the sidebar.

Now, to put people *into* your new group, click **All Contacts** at the top of the sidebar. Find the first person who belongs in this group (feel free to use the search box at the top), and drag that name directly onto the group's name. You're not moving that card out of the main list; you're just indicating that you also want it to appear in the group.

(Here's another way to make a group: Start by selecting a bunch of names in the Names column, ⌘-click them one by one, and then choose **File→New Group From Selection**. You get a new group that already contains all the selected names.)

If someone moves away, quits your company, or breaks up with you, you can remove them from a group. Click the group name and then the person's name; press the Delete key.

To remove this person from Contacts, confirm by clicking **Delete**. But if you just want to remove them from this *group*, click **Remove From Group**. ✦

---

edit, and then click **Edit**. Only now can you begin fiddling with this person's information.

Dictionary

The Mac makes it easy to look up words' definitions, figure out pronunciations, and settle arguments.

Type a word into the search box. As you go, the sidebar lists words that match what you've typed so far. Click one to read the definition and pronunciation.

Click **Thesaurus** to see synonyms and antonyms, **Apple** to consult an Apple glossary, or **Wikipedia** to check for an article on the topic.

> **TIP:** If you don't recognize a word in a definition, click *that* word to look up *its* definition. You can go down this rabbit hole forever.

Find My

It may appear as though the programmer who wrote this app fell asleep halfway through typing its name. Find My? Find My what?

As it turns out, what this app can find, on a map, is your lost iPhone or iPad—and your friends.

People

The **People** tab helps you locate your buddies as you try to find one another in the city, or keep tabs on little kids or older parents. It's not exactly a spy plane; you can't use it without the trackee's permission and awareness.

In fact, you have to let someone track *you* before you can track *them*. Suppose, for example, that you want to be able to see where your sister Sue is.

In the Find My app, click **Share My Location**, choose Sue's name, and click **Send**. Specify how long you want her to be able to find you.

Now Sue, wherever she is, gets a notification: "[Your name] started sharing location with you. Do you want to share yours?" She, too, can choose either **One Hour, End of Day,** or **Indefinitely** (or **Don't Share**).

For the time period specified, either of you can click the other's name in this app to see that person's position on a map. The ⓘ button offers options to

Contact this person, get **Directions** to her location, stop sharing your location with her, and so on.

Devices

Can't find your iPhone? Happens to everyone. Click its name on the **Devices** tab to see where it is on a map.

If it seems to be right where you are—meaning it's lost somewhere in your home—you can click the ⓘ and then **Play Sound**. For the next two minutes, your i-device pings loudly so you can figure out which room it's in.

Click to make the lost gadget play a sound.

The Find My app

If it's clear you left it somewhere in your travels—in a cab, in a restaurant—click the ⓘ and then, under Mark As Lost, click **Activate** and then **Continue**. Here's your chance to enter a phone number in case a good Samaritan finds the thing and wants to return it to you. Click **Next** and add a message ("$50 reward if you return this phone! Thank you!"). Click **Activate**.

Now, if someone picks up your phone and tries to wake it, your message and phone number are all they see.

> **NOTE:** Also on the ⓘ panel: **Erase This Device**. That's what you'd call the nuclear option; it erases the lost phone by remote control. In theory, you could use it if there are national secrets on your phone (or secrets that could get you into huge trouble).
>
> Realistically, of course, no spies can get into your phone without your fingerprint, faceprint, or password. But if it helps you sleep at night, erase away.

If Find My can't *find* your gadget—because its battery has died—you see its last known location. If it ever pops back online, you'll get an email. (Well, you will if you've clicked the ⓘ button and turned on **Notify When Found**.)

Font Book

From the day it was born in 1984, the Mac's talent with fonts (typefaces) set it apart. Today, Font Book is the app you use to install, remove, and look over your fonts. (The Mac comes with about 50 great-looking fonts, and thousands more are available to download and install. Some are free; some cost money.)

See the alphabet. See every character. Type a sample.

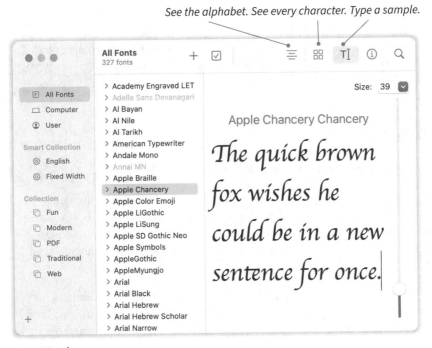

Font Book

So what can you do in Font Book?

- **Install a font.** When you download a font from the internet, it shows up as a double-clickable font file. When you double-click it, Font Book opens, shows you what the typeface looks like, and offers an **Install Font** button. (You may not see a newly installed font in your apps' Font menus until the *next* time you open them.)

- **Delete or hide a font.** To remove a font, click its name, press the Delete key, and confirm.

  More often, though, you may prefer to just *disable* a font. Click its name, click the checkbox above the list, and confirm by clicking **Disable**. The font no longer shows up in your apps—but it's still on your Mac, in case you ever want to use it again.

- **Spot duplicates.** If you see a dot next to a font's name, it means you've got more than one copy of it installed. Click the one you want to keep, and then choose **Edit→Resolve Duplicates**. Font Book turns off all *other* copies.

- **Print a sample sheet.** Click a font's name and then choose **File→Print**. In the Print dialog box, click the **Show Details** button. Now you can see the **Report Type** pop-up menu, which lets you choose from three different font-sample printout types. Click **Print**.

> **NOTE:** In Font Book's sidebar, you can see subsets of your fonts in "folders" called Collections. You can make your own collection, too; choose **File→New Collection** and type a name for it. Now click **All Fonts** and look for the fonts you want to add; drag them one by one onto the new collection icon.

Home

HomeKit is Apple's home-automation standard.

If you have enough money (and care enough), you can buy all kinds of home-automation products. These are "smart" or "connected" thermostats, door locks, doorbells, security cameras, power outlets, lightbulbs, sensors, and so on. You can control them all from your phone, even when you're not home.

Every major tech company has dreamed up its own software standard for controlling these things; Apple's is called HomeKit. If your "smart" product's box says "Works with HomeKit," you're in luck: This app, Home, can control it.

You have to use an iPhone or iPad to set up these gadgets, but once that's done, the Home app on the Mac can control them. You can turn them on or off, adjust their settings (turn down the thermostat, for example), change their schedules (such as when lights shut off), and so on.

The Home business gets complicated. You can set up various smart gadgets, and then collect them into software "rooms," and then put the "rooms" into "zones," and set up automated schedules and "scenes" for each one. Still reading? Here's Apple's complete HomeKit user guide: support.apple.com/en-us/HT204893.

Image Capture

This poor little app, born in 2002, has long since been leapfrogged in features and usefulness by Photos (for importing pictures from a camera) and Preview (for operating scanners). But it's still kicking around for the benefit of the Alliance of Image Capture Old-Timers.

Keynote

Keynote is Apple's version of Microsoft PowerPoint. That is, it's a slideshow app. It's an ideal app for making full-screen, slide-by-slide presentations of any kind: a talk, a class, a report, a proposal, an actual photo slideshow.

Map of your presentation   *Current slide*   *Formatting controls*

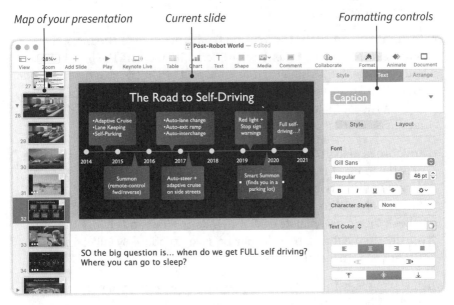

Keynote

As with Pages and Numbers—the other two apps in the iWork trio—Keynote starts you off with a selection of handsome, ready-to-go designs. Double-click one to open it.

Now build your slideshow:

- **Click to type over the starter text** (like "Presentation Title").

- **Click Add Slide to create the next screen.** You'll discover that a Keynote template isn't just one slide design; it's a selection of about 15 themati-cally unified designs. One is for the title slide. One is for showing a photo, or several. One has a placeholder for bulleted text points. One is just right for a quotation or a big statistic. And so on. Choose from this pop-up menu to add the slide, and then fill it in with your own text and graphics.

- **To insert a picture or video** from your Mac, choose Insert→Choose.

- **Use the formatting panel** at right to change the fonts, styles, line thicknesses, colors, and other aspects of whatever you've clicked. On the Arrange tab, for example, you can control which objects are "in front" of other ones or make things snap into alignment with each other.

- **Use the toolbar at top** to add tables, charts, text boxes, shapes, photos, or videos.

- **In the box below the slide,** type any notes you'll want to see, privately, while you're giving the talk. (If you don't see this box, choose View→Show Presenter Notes.)

- **To create an animated transition from one slide to the next,** click Animate (top right) and then Add an Effect. Choose from the list of effects, and then tweak the timing and other aspects right there in the right-side panel. You'll discover that Keynote's transition styles are clean, classy, and really cool-looking; none of them replicate the cheesiness that once gave presentation software a bad name.

- **Keynote is also great at *builds*,** adding another element each time you click the mouse (or press the space bar)—a useful technique for making bulleted talking points appear one at a time so your audience doesn't have the chance to read ahead. Select the text block containing the bullets; then click Animate→Build In→Add an Effect and choose the style you like. In the formatting sidebar, be sure to choose By Bullet from the Delivery pop-up menu.

When you're ready to practice your presentation—or even to present it—hit Play. The slide fills your screen, and the Mac thoughtfully prevents notifications from popping up while you're in front of an audience ("This is Rocko from Acme Collection Services. Your account is still overd ... ").

To advance the slides, click anywhere or tap the space bar. You can even jump around in the slideshow by typing, on your keyboard, the slide number you want (*1* to return to the beginning, for example).

If your laptop has a Touch Bar (page 73), you can tap the thumbnail images of your slides right on the bar—or **Exit** to stop presenting.

> **TIP:** If, at some point, you want to expound on a topic without a risk that your audience is distracted by your slide, tap the *B* on your keyboard. That blacks out the entire screen until you hit *B* again. (You can also hit *W* to make the screen all white instead.)

When you're done presenting, you can exit the full-screen presentation mode by tapping the Esc key on your keyboard.

Here's a great feature of Keynote: You can hide slides without removing them altogether. That comes in handy when, for example, someone asks you to give a 20-minute version of your usual 45-minute talk. In the sidebar, click the thumbnail of each slide that you don't want to appear (this time) and choose **Slide→Skip Slide**. You'll see its thumbnail collapse into a horizontal bar, meaning it won't appear when you play the slideshow.

Later, when you want to restore that slide to the lineup, you can click the collapsed bar and then choose the cheerfully named **Slide→Unskip Slide** command.

Maps

Maps is not quite as accurate as Google Maps, and there aren't quite as many features, but it certainly is pretty. Its purpose is not just to provide travel directions but also to give you information about any business in the civilized world—hours, phone numbers, website, reviews, photos, and so on.

Here's how to navigate (the app, that is, not the country):

- **Zoom in or out.** If you have a trackpad, pinch with two fingers to zoom out, spread two fingers to zoom in. You can also use the keyboard: Press ⌘-plus or ⌘-minus.

- **Scroll around the map** by dragging with the mouse or swiping with two fingers on the trackpad.

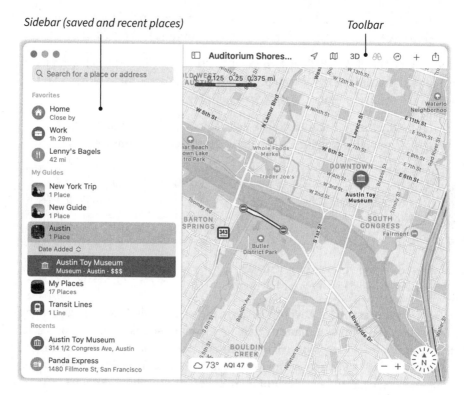

A tour of Maps

- **Rotate the map** by dragging the small compass at lower right or by twisting two fingers on your trackpad. (To return to north-is-up view, click the compass.)

- **Tilt into 3D view** by clicking **3D** at the top right of the screen. (3D view, which makes it look like you're looking down at the map at an angle, doesn't become available until you've zoomed in pretty close to a city. It's *really* useful once you can see the actual buildings.)

- **Switch into satellite view**—beautiful aerial photography—by clicking the ⬭ icon (top right) and then clicking **Satellite**. (Your other options here are **Default**—the standard, clean schematic of the world's roads—and **Transit**, which is great for inspecting train and subway lines.)

> **TIP:** The same little panel offers checkboxes for **Traffic** (which color-codes the lines of the streets, indicating how horrible the traffic is at the moment) and **Labels** (which lets you hide or show the street names in satellite view).

To pinpoint your own location on the map, click the ➹ at top right. A blue dot or a blue circle shows up. That's you.

> **NOTE:** Plenty of people don't appreciate having their locations tracked by apps. For that reason, the Mac circuit that detects your geographical location has an on/off switch. If it's turned off, Maps can't find you.
>
> To flip that switch, open **System Preferences→Security & Privacy**. Click 🔒, enter your Mac password, and turn on **Location Services→Enable Location Services**. (You can also turn Location Services on or off for each app individually in the list, for your paranoia-indulging pleasure.)

Searching Maps

When you click in the search box, before you've typed a single letter, Maps offers you one-click access to essential businesses you might be seeking right now, like **Fast Food**, **Gas Stations**, and **Restaurants**. (If you click **More** at the top of the list, it expands to include listings for nearby **Coffee Shops**, **Food Delivery**, **Groceries**, **Pharmacies**, **Pizza**, and other essentials.) When you click one of these categories, the sidebar lists nearby establishments in that category; click one to open its Details panel.

Categories appear before you type.   *Search by address or category.*

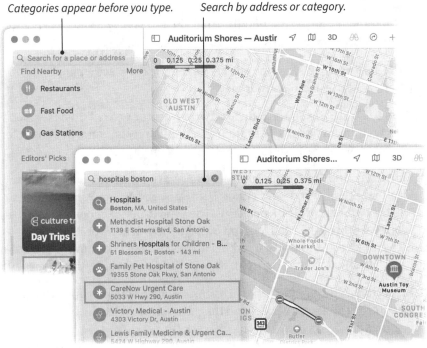

Searching

The starter list also includes Editors' Picks: nearby parks, shops, and other public attractions that Maps hopes might catch your eye. This feature, new in Big Sur, draws its information from popular travel guides like Lonely Planet and All Trails.

If any of that is what you're looking for, great. The rest of the time, you'll probably want to use the search box in the more traditional way. You can type in just about anything: a street address (*200 W 79 st, new york, ny*), an intersection (*s wabash and roosevelt, chicago*), a city, a ZIP code, a point of interest (*mount rushmore*), a kind of business (*emergency rooms memphis*), and so on. You don't have to capitalize anything, and you can use shorthand.

> **TIP:** Once again, Siri can make things a lot faster and easier. You can command her to "Give me directions to the airport," "Find Chinese food near me," "Zoom in on the map," "Mail these directions to Stacy," and so on. Your hands can remain lazily at your sides—or wrapped around your beverage.

As you type, Maps displays a menu of places that match. Click one to view its Details panel.

The Details Panel

Whenever you click a place—or the red pushpin button that represents one on the map—Maps opens the Details panel. It lists every conceivable action you might want to take at this point. For example:

- **Call (☏).** For a business, one click does the trick to place a phone call using FaceTime audio on your Mac.

- **Save to Guide.** Big Sur introduces this new feature, called Guides, which you can think of as a saved search: "Best Italian restaurants in Denver" or "Places to see in San Francisco." See the box on page 270.

- **Share (⬆).** The **Share** command lets you send a little bookmark for this location, either to another person (**Mail, Messages, AirDrop**) or to another app on your Mac (**Notes, Reminders**). The **Send to Device** command lists whatever other Apple gadgets you own, like your iPhone or iPad; the idea is that you've just looked something up on your Mac, but now you're about to go out, and it would be helpful to have the directions on the phone.

- **Directions.** The directions panel opens, ready to go, with your current location as the starting point and the place you've selected as the destination.

Click a place to read key details. *Scroll to expand the panel.*

Place details

- **Create Route** is practically the same thing, except that it lists this location as the *starting* point.

- **Look Around.** Apple has spent years driving the roads of big cities in specially equipped camera cars. The goal: to build up a massive database of *photos* of every street in every major city. If a **Look Around** tile appears on the Details panel, click it to view a photo of the address you've been studying. It's a great way to get a feel for the neighborhood—and for the classiness of that restaurant or whatever—before you commit to going there. (See page 272 for more on Look Around.)

- **Hours, Address, Phone, Useful to Know, What People Say.** If you're looking up a business, these informational bits serve as a giant global digital Yellow Pages—with Yelp.com thrown in for customer reviews.

- **Add to Favorites.** *Favorites* are bookmarks for places you go often. The dance center where your kid takes ballet, your best friend's house, the schools, your branch offices, whatever.

To add a place to your list, scroll down the panel and click **Add to Favorites**. From now on, it appears at the top of the sidebar list before you've even clicked into the search bar. (At that point, the button changes to say **Remove from Favorites**.)

> **TIP:** **Home** and **Work** are special kinds of Favorites—the two addresses that presumably you use most often. Click the ⓘ button to see what Maps thinks your home or work address is—and if it's wrong or missing, click **Open My Card** to edit your address card in Contacts.

MEET YOUR TRIP-PLANNING GUIDES

In macOS Big Sur, Apple introduces Guides: collections of locations you assemble for your own purposes.

Imagine that you're researching a trip to San Francisco. Every time you find a promising restaurant, museum, or point of interest, you can add it to this research collection, called a guide, for easy retrieval later when you're planning your day.

Or, if you're traveling to a conference, your guide might include your hotel, the conference venue, the restaurants you've booked, and the local airport.

To create a new guide, click the ● icon at the top right of the Details panel. From the menu, choose **Save to Guide→New Guide**. In the sidebar at left, a **New Guide** appears; ready for you to rename.

Now all you have to do is start adding the places you're researching to this guide. That's easy enough: On a place's Details panel, click the ● icon and choose **Save to Guide→San Francisco Trip** (or whatever you've named your guide).

You'll soon discover that your guide, in the sidebar, acts like a folder; the places you've added to it appear in the list indented. Yes, they're still basically bookmarks, but now they're bookmarks you can organize. ✦

- **Create New Contact** creates an address book "card" for this place in your Contact app.

- **Report an Issue.** Behind the scenes, the Maps app and its massive database of businesses are kept updated by humans. If you discover that the information about a place is wrong—if its hours have changed, say, or if it's gone out of business—click here to let Apple know.

- ⬤. Hiding at the top right of the panel is a single menu that lists most of the important stuff you've just read about: **Share, Call, Send to Device, Add to Favorites,** and so on.

The Directions Panel

Maps assumes you'll often want to use it for getting directions to a place. That's why **Directions** buttons are in your face at every turn.

When you click one, the directions panel appears. The two text boxes let you specify your starting point and ending point; they're usually filled in automatically. (The ⥮ icon swaps those two locations, which is perfect for planning your return trip.)

Maps proposes alternate routes.   *Click one to view the steps.*

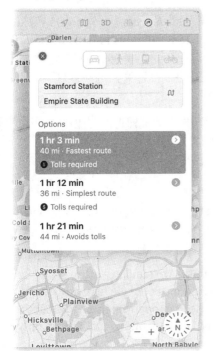
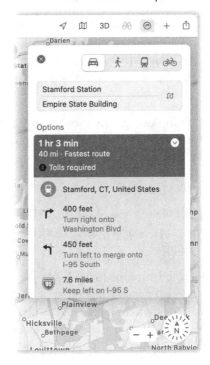

The directions panel

At the top, four icons let you indicate how you'd like to travel:

- **By car.** The list offers you several different routes, organized by driving time. The ● symbol marks toll roads.

  Double-click one of these routes to reveal the actual turn-by-turn directions. What's cool is that you can click each step of the directions to view that segment on the map.

  > **TIP:** Click **Options** to reveal a panel that lets you omit routes that involve paying tolls or driving on highways.

- **On foot.** Here again you can double-click a proposed route to see a list of the directions. In most cases, walking takes longer than driving, although in Los Angeles, that's not a sure thing.

- **Using public transportation.** Impressively enough, Maps knows about every bus, train, and subway in every major city, complete with its schedule. That's how it's able to build complete itineraries that even include the distances you'll have to walk between, for example, bus and subway. You can click the name of a train or subway segment to view the actual stops you'll be passing by and how many minutes are between each one.

- **By bicycle.** In Big Sur, for the first time, you can choose to travel by bike. The step-by-step directions include the kinds of information you'd want to know, like how steep each segment is, when you'll have to carry your bike up stairs, and what kind of road you're on (bike lane, main road, side road).

Don't forget that once you've got a route chosen, you can hit ⛴ at the top right of the window to send it to your phone, to another person, or to your Notes or Reminders apps.

Look Around

Street View is one of the most famous features of the world's most popular maps app, Google Maps. It's a 360-degree photographic representation of any address.

Not to be outdone (usually), Apple created its own similar feature called Look Around. It's faster, smoother, and sharper than the Google feature; when you click to move from one spot to another, it seems to morph into video mode to fly you through the streets. Unfortunately, Look Around is available in only about 15 cities.

To try it out, search for an address in a major city like Chicago, Seattle, New York, Las Vegas, San Francisco, Boston, or Los Angeles. On its Details panel, click **Look Around.**

Suddenly, a photo appears: what you would see if you were in that city, standing in the middle of the street. To look around you, drag the mouse horizontally. To move down the street, click a more distant spot. You can tour a whole neighborhood like this, without paying a nickel in gas.

When you've finished looking around, click ❌. And send waves of encouragement to Apple to add the rest of the world's 20,000 biggest cities.

> **TIP:** In macOS Big Sur, Maps can even show you maps of the *interiors* of 115 major airports and 375 shopping malls. (Make sure you've zoomed in close enough, and make sure you're not in satellite view.) You can see the actual store and restaurant names and whether they're before or after airport security—information that could be worth its weight in gold.

AERIAL CITY TOURS WITH FLYOVER

Apple hasn't just been driving the planet with camera cars. It's also been flying camera-equipped helicopters.

As a result, Maps allows you to conduct your own aerial tour. The trick is to search for one of the 350 major world cities that Apple has photographed from its helicopters. Switch to satellite view, turn on **3D** (top right of the window), zoom in until you can see the individual buildings, and marvel at your self-guided virtual aerial tour of the city.

Move the map by dragging with your mouse (or sliding two fingers on the trackpad). Change the camera angle by dragging the tiny slider at lower right (or by Option-dragging with two fingers on the trackpad). Zoom in or out with the − and + buttons at lower right (or by pinching or spreading two fingers on the trackpad).

If that's too much work, you can let Maps drive the tour. This feature is called Flyover Tours, and it's pretty spectacular.

To see one, search for a major city—say, San Francisco. On its Details panel, click **Flyover Tour**. Sit back in a slack-jawed trance and watch as Maps displays a gorgeous, preprogrammed aerial tour of that city. ✈

Notes

Notes has grown, over the years, into a powerhouse of a data gobbler. It can store notes of any kind: brainstorms, recipes, phone numbers, driving directions, credit card numbers, frequent flyer numbers, and so on. But it can also store photos, documents, tables, checklists, videos, maps, web links, and just about anything else you might want to retrieve later.

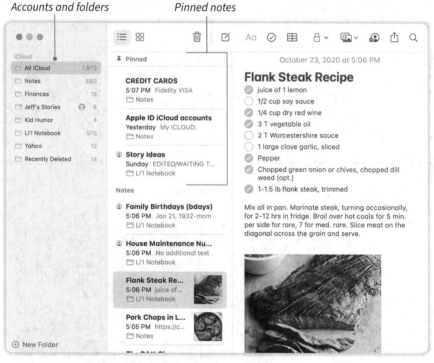

The Notes app

Three Panels

In its fully expanded condition, Notes offers three window segments:

- **Folder list.** You can organize your notes into folders, whose names appear here. (Click **New Folder** to create one, name it, and then drag individual notes into the newly created folder in the list.)

- **Notes list.** Here you see the title and first line or so of every note, along with the time and date you last edited it, for quick table-of-contents-ish scanning.

Your notes are listed chronologically, with one exception: At the top of the list are *pinned* notes—the most important ones that you want to refer to frequently. And how did they get here? You put them here, by selecting each one and then choosing **File→Pin Note**.

- **The Note itself** fills the main window.

Making a Note

To add a note, choose **File→New Note,** or click ☑ on the toolbar. You can type, drag, or paste just about anything into the new note.

> **TIP:** The first line of text becomes the note's title. If you like, Notes can automatically make it stand out in big, bold type—whatever style you choose in the **Notes→Preferences→New notes start with** pop-up menu.

It's fine to just type plain old text into a note. But don't miss some of your other options:

- **Paragraph styles.** Using the Aa menu on the toolbar, you can apply a variety of headings to any selected paragraph. Or you can turn some selected paragraphs into a bulleted list (much like the one you're reading now), or a numbered list, or a list that uses dashes instead of bullets. This menu also contains buttons for **bold**, *italic*, <u>underline</u>, and ~~strikethrough~~.

- **Checklist style (⊘).** This button turns any selected paragraphs into a handy to-do list. Click in a circle to produce a green checkmark, which means "done."

> **TIP:** The first time you use this feature, Notes offers you an **Enable Sorting** button. It means when you check off an item, it slides to the bottom of the list, thereby calling less attention to itself. (You can also turn this feature on or off in **Notes→Preferences**.) Notes is all about keeping things tidy.

- **Table.** When you click the ⊞ button on the toolbar, you get a 2 × 2 grid: a tiny spreadsheet. To add columns or rows, click inside a cell; click one of the little table handles (⊡ or ⊡); and then click the command menu (∨) to produce the **Add** and **Delete** commands for rows and columns.

> **TIP:** When you're typing in the lower-right cell, you can also make a new row by pressing Return.

At any time, you can convert the table into regular text by selecting the rows or columns in question, clicking the ⊞ button on the toolbar, and then clicking **Convert to Text**.

What you can't do, alas, is adjust the column widths or row heights. What do you think this is, Microsoft Excel?

- **Add a graphic.** *Continuity* is the ingenious Apple suite of features that connects an iPhone to your Mac wirelessly—and the 🖼 on the toolbar is a perfect example.

This menu offers three options. **Take Photo** lets you use the camera on your *iPhone*, if you have one, to take a picture—which then appears here, on your Mac! **Scan Documents** also powers up your iPhone's camera, but this time it opens a special mode that's designed to scan printed pages, automatically straightening and sharpening them. And **Add Sketch** produces a blank screen on your iPhone; you can use the pen, pencil, and highlighter tools to make a little drawing with your finger. As soon as you hit **Done**, that image appears in the note on your Mac. It's very cool.

Sharing a Note

Notes is also a *collaborative* app: You and selected special friends can access the same note simultaneously over the internet, editing or consulting the same page. This is a perfect feature for brainstorming together, working on a party guest list, sharing a recipe, or whatever.

The sharing starts here.

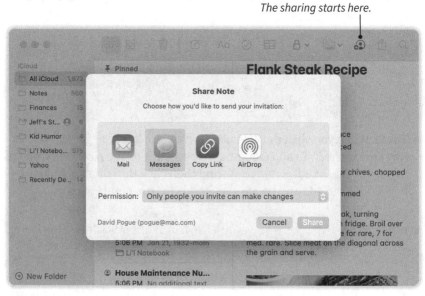

Sharing a note

To begin, click the 👥 button on the toolbar. The Share Note screen appears. Now you're supposed to choose how you would like to invite your co-conspirators to this note: by email, text message, or AirDrop, or by copying an internet link, which you can then paste into any program you like.

Use the **Permission** pop-up menu to specify whether these people will be able to just *look at* your note or make changes, too.

Finally, click **Share**. If you chose **Mail** or **Messages**, here's where you say *who* you're inviting.

Once your collaborators have received and accepted your invitation, this note, or folder full of notes, appears in their Notes apps, and they're free to edit (or just look at) the notes you've shared. At any time, if somebody misbehaves, you can rescind permission by clicking the 👥 icon on the toolbar and adjusting the sharing settings.

Locking a Note

Sometimes the information on one of your notes pages is best kept private. Maybe it's financial information. Maybe it's a planning list for a surprise party. Maybe it's something that's not safe for work (or family).

Fortunately, you can password-protect a note. Click the 🔒 on the toolbar, choose **Lock Note**, make up a password, and click **OK**. From now on, this note reveals nothing until you enter the password.

Numbers

Next time you need to enter data in rows and columns, keep in mind that your Mac came with a spreadsheet program. It's not as full-featured as Microsoft Excel, but it's not nearly as complicated and bloated, either. It has more than enough flexibility to pop out handsome lists, charts, and graphs.

When you open the app, Numbers offers you a screen full of *templates*—ready-made spreadsheets with all the auto-calculation programming built in—that you can tweak to taste and then fill with your own information.

These starter spreadsheets are excellent for checklists, making charts, creating budgets, tracking your savings, tracking stocks, tallying your net worth, calculating your grade-point average, comparing loans, recording your baby's milestones, planning a meal or party or trip, organizing a team, and so on.

If you think one of these will do the trick, just double-click it. The templates come with placeholder text and numbers in the individual cells; just double-click a cell to edit it.

If not, you can always design your own spreadsheet by opening the one called **Blank**.

YouTube is full of tutorials on using spreadsheet apps, but the basics are straightforward: You can type or paste anything into each cell of the table. You can set up formulas in certain cells that perform math on other cells (for example, adding them all up; use the **Insert** pop-up menu to choose the operation you have in mind).

Once you've got a table full of numbers, you can highlight it and then use the **Chart** menu to represent it as a chart or graph.

Because it's an Apple app, Numbers makes it really easy to put graphics and even videos into cells. That makes it a natural for organizing, for example, headshots in a team directory, an inventory of valuable stuff in your home, or a table of what you've planted in your garden.

Pages

Word processing may be the single most common computer productivity task, so it's no surprise that Apple includes Pages with every Mac. You can use it as a word processor, just like Microsoft Word; in fact, it can open and create Word documents, so your colleagues never need to know you haven't rendered money unto Microsoft. But it's also a page-layout program, so you can make your documents look a little bit fancier than typed words on a sheet of paper.

As with Numbers, Pages starts you out with a few screens full of predesigned document templates, for making résumés, term papers, business reports, correspondence, flyers, newsletters, certificates, and even full-length novels. If you like the looks of one, double-click to open it. Each text block starts out full of bogus placeholder text; just click there and then start typing over it.

You can liven up a wash of visually boring text by inserting pictures, charts, tables, and so on, using the toolbar at the top. Then you can format each of those elements using the **Format** sidebar at the right side: Add a border to a

Format and inspect whatever is selected.

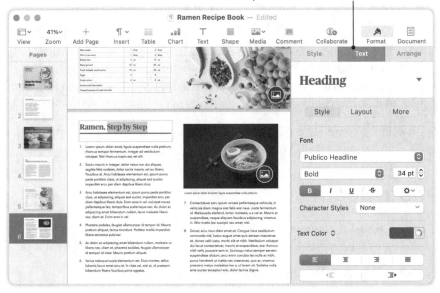

Pages

picture, control whether or not text flows around the image, adjust the color tone of an image, and so on.

There are, of course, spelling and grammar checkers (**Edit→Spelling and Grammar→Check Now**), an option to collaborate with other authors across the internet (**Share→Collaborate with Others**), and revision tracking, which color-codes the changes people make (including you) so you can see the proposed edits before committing to them (**Edit→Track Changes**).

When you're finished writing and designing your masterpiece, you can save it as a document to share with other people. The **File→Export To** menu offers choices like these:

- **PDF.** Yes, you can spin out your document as a PDF file, the universal exchange format that virtually any computer on earth can open. The recipients will see it with precisely the graphics, layout, and font choices that you created on your end. (The downside—or the upside, depending on your point of view—is that PDF documents aren't easy to edit.)

- **Word** means, of course, Microsoft Word, the world's most popular word processing format. You can send this kind of document to anybody who uses Microsoft Office, confident that they'll be able to open it, read it, and make changes to it. (In this case, however, it may not look exactly the way it did on your screen, because their computer might not have the same fonts yours does.)

- **EPUB** is a standard, not-copy-protected electronic book format. People who have the Kindle ebook reader, for example, can open and read your work if you supply it in this format.

Also in the **File** menu: **Publish to Apple Books**. That's right: You can actually publish something you've written using this command. It shows up in the Apple Book Store, right beside the bestsellers from J.K. Rowling and Stephen King. And if anybody buys it, you can actually make money from it.

Apple offers a complete guide to self-publishing this way at https:/support. apple.com/en-us/HT208716.

Photo Booth

It's aging, it's goofy, it's very nearly pointless—but Photo Booth offers a quick way to take pictures or videos of yourself, using your Mac's built-in camera.

To begin, click one of the three tiny icons at lower left:

- **4-up** is the closest you'll get to an actual photo booth in a shopping mall. It snaps four consecutive pictures in two seconds, giving you enough time to make four different expressions and poses.

- **Still picture** snaps a single photo.

- **Movie clip,** of course, shoots a little video.

To begin the capture, click the red ⊚ button or press Return. The app makes a 3-2-1 countdown, and then your Mac shoots the shot or the video. (To end the video, hit the ⊙ button.)

The new photo or video appears in the row of thumbnails at the bottom of the window. Click one to examine it; if you like, click the ⛓ button to send it off to your admirers.

> **TIP:** Like a mirror, Photo Booth reverses your image horizontally. Writing on your shirt appears backward in the picture, and your face and hair are flipped left to right.
>
> You can "fix" a flipped photo by selecting it and then choosing **Edit→Flip Photo**. Or you can choose **Edit→Auto Flip New Items** to make Photo Booth give up its silly mirror impersonation altogether.

Photo Booth's real-time distortions are hilarious, especially if you never grew up.

4-up, still picture, movie clip   Begin countdown

Photo Booth

Effects

Without a doubt, the best part of Photo Booth is the **Effects** button. It opens three screens full of filters that perform various hilarious distortions to your image. They make your eyes bug out or your mouth swell up, or they split the screen down the middle so you become a conjoined twin with yourself. Some animate hearts or flapping bluebirds over your head—and they actually follow you as you move. Others play with the color, turn you into an X-ray, mimic a thermal camera, and so on.

If your goal is to entertain someone whose emotional age is between 4 and 14, you'd be hard-pressed to find anything more addicting than Photo Booth and its effects.

Photos

If you're like most people, you have a smartphone. If you're like most people with smartphones, you take a lot of pictures. And if you take a lot of pictures, you need some way to organize them.

That's what Photos is all about. It's a massive digital shoebox for collecting, editing, organizing, and sharing the photos and videos of your life. If you've ever used Photos on the iPhone or iPad, you'll feel right at home; the Mac version is almost identical.

Most of the time, you'll be looking at your pictures in the form of thumbnails, small enough that you can view a whole bunch simultaneously. (There's a slider at top left that lets you make the thumbnails bigger or smaller.)

But when you want to open a photo at full size, or as big as your screen will allow, double-click it.

> **TIP:** If you've selected a thumbnail by clicking it, you can press the space bar as a shortcut to opening it. When you've had a good look, press the space bar again to return to the thumbnails.

Sidebar (categories and albums)   *Toolbar*

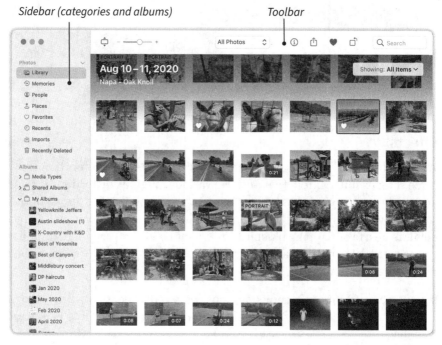

The Photos app

At this point, you can use the controls along the toolbar at the top right: Click ⓘ to read the photographic details of the shot, ⬆ to send it to somebody, ♡ to mark it as a favorite (it appears in the Favorites "folder" in the sidebar list), ⬜ to turn the image 90 degrees, ⚡ to beautifully color-correct it, or **Edit** to enter the Mac's Photoshop mode, described in a moment. You can also magnify the photo even more, using the zoom slider at the left end of the toolbar.

If you'd like a tour of Photos, consider reading down the sidebar at the left side of the screen. Its headings form a nice map of the app's features.

Photos

Click here when you want to see all your photos at once, represented as thumbnails.

You can scroll down through them all, going back to the first picture you ever took—the oldest appear at the top—or you can look at them in groupings by **Years**, **Months**, or **Days** by clicking the tabs at the top.

> **TIP:** If **View→Metadata→Titles** is turned on, you see each photo's name beneath its thumbnail. Unfortunately, they usually look like *DSC_8823.HEIC* or *IMG_4203.jpg*. Fortunately, providing more descriptive titles is easy: Click the existing name to open its renaming box, and type away. Later, you'll be able to find these photos by typing a keyword or two into the search box—grateful that, in Big Sur, these captions sync to your iPhone and other Apple machines.

Memories

Apple calls them Memories; you might call them "musically accompanied slideshows, created by artificial intelligence, based on commonality of time, place, or subject."

And sure enough: When you click **Memories** in the sidebar, you find miniature billboards of ready-to-play slideshows with names like "San Francisco Trip," "On This Day Last Year," "Grand Canyon," "Dad Over the Years," and so on.

Photos automatically creates about three new Memories every day, which does a great job of resurfacing shots that you would otherwise never look at again.

When you double-click a memory, its Details page opens. What you'll do here usually is just play the darned thing; you get a beautiful, animated, sweetly crossfading slideshow, accompanied by glorious music.

Click to begin the video.

During playback, click to adjust length and soundtrack.

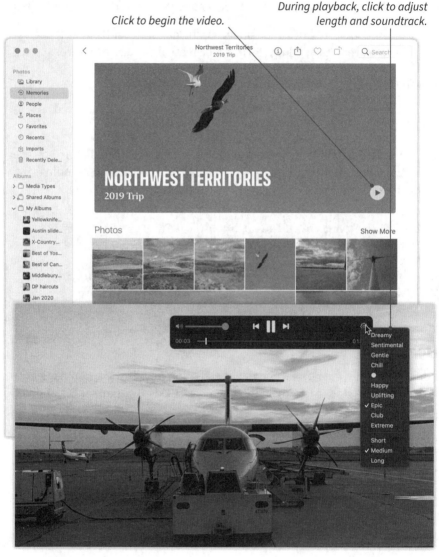

Memories

People

Photos can actually tell who is *in* your photos. We're talking facial recognition that's good enough to recognize the same child at ages 4, 10, and 16.

It does, however, need a little assistance from you to get started.

First, Photos needs a little private time to analyze your photo collection. If it's a big collection, this could take a day or two.

Click **People** in the sidebar. Now you're seeing the thumbnails of the faces Photos has found so far. Click beneath a thumbnail to enter that person's name, so Photos will know who it is later.

Faces you drag to the top become Favorites; they get bigger headshots.

Double-click a headshot to see all photos of that person.

The People screen

And here's the best part: You can double-click one of these faces to see all the photos of this person in your entire Photos collection. You'll be really grateful for this feature the next time somebody asks for a couple of photos of you (or anyone in your collection).

> **NOTE:** Every now and then, Photos announces, "There are additional photos to review." It's telling you it's tried to identify the people in some new photos but needs you to confirm. Each time you do that, Photos gets better at recognizing your friends and family.

THE MIRACLE OF ICLOUD PHOTOS

The service called iCloud Photos may be one of Apple's greatest inventions. If you turn it on, you give permission for Photos to store your entire video and photo collection online, on Apple's servers.

And why is that such a miracle? Because it means every computer, phone, or tablet you own is always up to date with all your photos, photo albums, *and* whatever editing and organizing you've applied to them. You never have to wonder: "Was that picture on my Mac or on my iPhone?" They're all everywhere.

It also means that the most precious files you own—your photos and videos—are always backed up, wirelessly and in real time. Your house could fall into a sinkhole, engulfed in flames, and you wouldn't care. You'd know that your photos were still safely stored online. (OK, you might care somewhat.)

iCloud Photos also means you have access to your entire photo collection from any computer or phone in the world. You just log in at iCloud.com, click **Photos**, and there they are.

Maybe most useful of all, using iCloud Photos can grant you a Mac whose hard drive never gets full. If you turn on a feature called **Optimize Mac Storage**,

then whenever your hard drive does start getting full, Photos quietly removes the huge, space-eating, full-size originals of your photos from the Mac—and replaces them with scaled-down copies that occupy far less disk space. When you're working in Photos, you probably won't even know the difference, because you can still work with them, edit them, organize them, and zoom into them until they fill your screen.

Then, if you ever want to export, print, or share those photos, Photos instantly downloads the full-resolution originals, quickly and invisibly, for you to work with.

All these benefits may sound joyous and effortless, and they are, but beware: They cost money. Apple gives you 5 GB of storage space with your free iCloud account, which is nice. But unless your photo collection is very small, 5 GB is not enough to hold it. You'll have to upgrade your iCloud storage. Apple charges $1 a month for 50 gigabytes, $3 a month for 200 GB, or $10 a month for 2 *terabytes*. (Unless you take so many photos that your family hates you on vacations, the $1 plan is probably plenty.)

If you decide to try iCloud Photos, choose **Photos→Preferences→iCloud.** Turn on **iCloud Photos**. ✦

Places

Click this button in the sidebar to see a map, dotted with photo clusters taken in each location. Double-click over and over again, zooming in each time, until you're looking at the actual photos taken in that spot.

> **NOTE:** Smartphones automatically tag each picture with its geographical coordinates. Traditional cameras, however, usually don't, so those pictures don't show up in Places.

Recents, Imports

Recents are all the photos that have entered your life electronically lately— from your phone and from other people. And **Imports** shows every batch of photos you've ever brought into Photos, chronologically, scrolling back to the dawn of photographic time.

Recently Deleted

Photos has its own little trash, and it exists for exactly the same reason as the one in the Finder: to save you from yourself. Any photo or video you drop into it self-destructs, but not for 30 days.

During that second-chance interval, you can open **Recently Deleted**, click the doomed photo, and then click **Recover**. Photos puts the photo back wherever it came from originally.

Media Types

These folders let you look over your photographic exploits according to what *kind* of file they are. Most of them assume that you have an iPhone, because the file types correspond to the different modes in its Camera app: **Selfies, Live Photos, Portraits, Panoramas, Time-Lapse, Slo-Mo,** and so on.

Activity, Shared Albums

When you get right down to it, there are really only two reasons people take pictures. One, of course, is to capture a moment so you can recall it later. But the other is for sharing, to show *other* people what you experienced.

The most obvious way to send a photo to somebody else is to email it or send it as a text message. Click to select the photo (or ⌘-click several), and click the ☐ near the top right of the window. Choose **Mail** or **Messages,** and off you go.

The trouble with that method is that it sends scaled-down images, not the full-size originals. It's also a one-way process. It offers no *interaction* with the recipients; they can't give you a thumbs-up, leave comments, or contribute their own photos.

That's why Photos also offers **Shared Albums**. Using this method, your audience can respond to your shots the way they would to photos on Facebook or Instagram: They can click a **Like** button, leave comments, and even contribute their own pictures.

> **NOTE:** If you don't see **Activity** and **Shared Albums** in the sidebar, it's because you haven't yet turned on the feature. Open **Photos→Preferences→iCloud**; turn on **Shared Albums**.

Here's the process:

1. **Round up the pictures and videos.**

   For example, you might put them into an album (described in the next section). Or, to select a random bunch of thumbnails, click each while pressing the ⌘ key.

2. **Create the shared album.**

   Click ⬆️→**Shared Albums**. When the Add to Shared Album sheet appears, type in a Comment, if you like, and then click **New Shared Album**.

1. Choose **Shared Albums**.   *2. Create the album.*   *3. Invite the audience.*

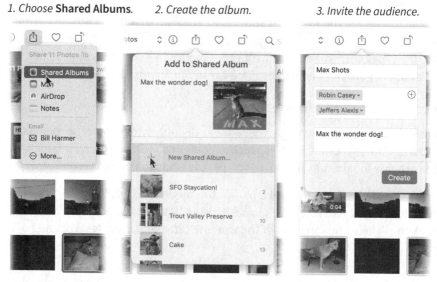

Making a shared album

3. **Name your album.**

   Type the name other people will see into the **Shared Album Name** box.

4. **Invite your audience.**

   In the Invite People box, click the ✚ button to view a miniature copy of your address book. Choose the email address or phone numbers of each person you're going to invite to see your photos.

   In general, only members of the Apple ecosystem—people who have their own iPhones, iPads, or Macs—get the full Shared Album treatment, and only their iCloud email addresses and phone numbers show up here.

   > **TIP:** You *can* still share your photos with unfortunates who don't have Apple devices; read on.

5. **Click Create.**

   At this point, everybody you invited gets a notification in their Photos apps (iPhone, iPad, Mac) that you shared an album. And you see the name of the shared album in the **Activity** "folder" in your sidebar, so you can have a look at what you shared.

If you click the name of your shared album, you can make changes like this:

- **Add or remove photos and videos.** Any photos you delete here disappear from everybody else's machines, and any you add (click **Add photos and videos**) show up on their screens within a few seconds.

- **Prevent others from contributing.** If you and some of your audience were at the same party, concert, or trip, they can contribute some of their *own* photos to your shared album. Pretty cool: new angles, new subjects.

  Unless, of course, you fear either mischief or a sullying of your artistic vision. In that case, click the ☺ icon at the top right of the window and turn off **Subscribers Can Post**.

- **Post the album on the web.** If you click the ☺ and turn on **Public Website**, then even people not blessed with Apple equipment can look at your pictures. "Public" here does not mean the teeming masses will have access to your fine art; it means only that you can now add the email addresses of non-iCloud people. Their invitations will include a web link that opens up a private web page featuring your magnificent photography. (They won't be able to add comments or contribute their own photos. That's what they get for being second-class ecosystem citizens.)

- **Rethink the invitees.** That ⊕ icon also reopens the addressee box, where you can invite new people, or delete old people, from the invite list.

Albums

In Photos, an album is just what you'd expect: the digital version of a printed photo album, meaning you've hand-selected which photos appear there and in what order.

But the electronic versions have certain advantages. For example, the same photo can appear in a bunch of different albums without actually being duplicated and thereby taking up more disk space. You can remove a photo from an album at any time, confident that you're not actually removing it from *Photos*—only from this virtual grouping.

To create a new album, choose **File→New Album**. It appears in the My Albums section of the sidebar with the temporary name **Untitled Album**; type a better name for it.

At this point, you can explore the other headings in the sidebar—**Photos, Favorites, People**, or whatever—and drag thumbnails onto the name of your new album to install them there.

My Projects

Hard though it may be to believe, some people still like to look at pictures on paper. Fortunately, for a small fee, plenty of companies can turn your photos into gorgeous calendars, prints, cards, wall posters, and photo books. Your project gets professionally printed on shiny paper, handsomely bound, and shipped to you or your adoring fans.

Apple has integrated these companies' publishing options with Photos—sort of. When you choose **File→Create→[project type]→App Store**, the App Store screen opens before you. It lists a bunch of *extensions* for Photos—plug-ins, each adding a new feature to the app, like the ability to design cards, books, calendars, and so on, and order them from the corresponding printing company.

The next time you open Photos, you'll see that company's name in the **File→Create→[project type]** submenu. Click through enough introductory screens and you eventually arrive at a layout window, where you can choose the photos you want to include in your project, design it, and order it for shipping.

Thereafter, you'll be able to revisit the project you edited by clicking **My Projects** in the sidebar.

Editing Photos and Videos

Over the years, successive versions of Photos have crept closer and closer in power to professional touch-up programs like Photoshop. And in macOS, you can apply most of those same changes—rotation, cropping, brightness, saturation, contrast, and so on—to *videos* as well as photos.

Here's the best part: *No editing you do here is ever permanent.* It's all non-destructive editing. That is, you can return to a photo, minutes or even years later, and restore it to its original, unedited state.

THE RETOUCH AND RED-EYE TOOLS

Most of the adjustment controls in the editing mode of Photos work the same way: Play around with the sliders until you like what you see. Each one affects the *entire* photo.

But two of the tools let you actually *click* the photo to indicate where you want them to operate.

With the **Retouch** tool, for example, you're supposed to click spots in the photo that are marred by irregularities: a hair, a speck of dust, a pimple, a scratch, or a streak on an old photo.

First, turn the tool on by clicking the ✐ icon. Then, using the **Size** slider (or by pressing the [and] keys), adjust your cursor size until it's just bigger than the blemish. Finally, click directly on it in the photo. Photos magically eliminates

the problem, filling in the spot with an ingenious blending of pixels from the surrounding area.

The other direct-painting tool is **Red-Eye**. It's designed for those rare cases when the camera's flash illuminated the pupils of your subject, giving them a devilish red glow. Click the ✐, adjust the **Size** slider, and then click directly inside each eye that exhibits the problem. Photos replaces the red with black. ✦

How do you enter editing mode? If you're looking at your photos in a thumbnail view, click one and then press the Return key. Or, if you've opened a photo—it's filling your window—press Return or click **Edit** at top right.

At this point, the Photos editing window feels like a separate app. It has its own layout and a totally different look.

At the top right, for example, Apple has equipped you with the two photo-repair tools that most people use most often: **Rotate** (⌑), which turns the photo or video 90 degrees with each click, and **Auto Enhance** (⚡), which instantly fixes the color, brightness, and contrast of your photo or video according to a mountain of artificial intelligence. Usually, it does a great job of improving the look of a photo or video; sometimes, it doesn't make much difference; occasionally, you'll wince and immediately click the ⚡ again to turn off the "enhancement."

On the **Adjust** tab, the "sidebar" on the right side lists the 13 primary areas of repair work you can do to a photo: **Light, Color, Black & White, Retouch, Red-eye, White Balance**, and so on.

Under **Light**, for example, you get **Brilliance, Exposure, Highlights, Shadows, Brightness, Contrast**, and **Black Point**.

Each of these adjustments is affected by the settings of the component sliders below it. (If you don't see the component sliders, click the ⟩ next to the heading—**Light** or **Color**, for example—to expand the controls; click ⌄ to collapse them. And why did Apple make these headings collapsible like this? Because if they were all expanded all the time, you'd need a monitor 11 feet tall.)

For each major heading, you can make adjustments automatically, semiautomatically, or manually:

- **Fully automatic.** As you point to each heading, an **AUTO** button appears. It means "Photos will adjust all the sliders automatically to the best of its little software ability."

  Often, that's the only fix you need to make, and you've just saved the time it would've taken you to fiddle with the individual component sliders.

- **Semiautomatic.** Once you've expanded the heading, the first thing you see is an inch-tall "filmstrip" of the photo, showing a gradation of light or color changes. As you click or drag inside this strip, Photos operates all the component sliders simultaneously. It's not as one-click as the **AUTO** button, but at least you're adjusting multiple parameters with only a single cursor move.

- **Manual.** Once you've expanded an adjustment heading, you can operate the underlying sliders in any of three ways: You can click inside the slider to move its setting point. You can drag, watching the changes to the image as you make them. And you can *double-click* anywhere in the slider to return its setting to the original zero point.

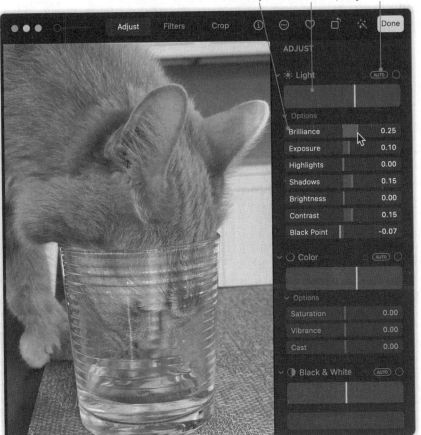

Manual, semiautomatic, fully automatic

The editing window

In each case, the little numbers off to the right give you a numeric sense of what setting you've applied. You might want to keep them in mind when you attempt to edit another photo the same way.

Among these 43,000 sliders, here are some of the most useful:

- **Brilliance** brightens the dark areas and slightly dims the brightest spots, revealing details that were muddy or blown out. (You can also perform these two tasks independently: De-muddy the darkest areas using the **Shadows** slider, or de-blowout the highlights using the **Highlights** slider.)

- **Saturation** intensifies all the colors. If you apply it modestly, you add a little color pop to a slightly washed-out shot; if you apply it heavily, you can turn a landscape into a crazy neon psychedelic poster.

- **Black & White** turns a color photo into an artsy grayscale image.

- **Curves and Levels** are advanced controls that should look familiar if you've ever played around in Photoshop. They provide individual *histograms* (graphs) of the color distribution—reds, greens, blues—in the photo.

  By manipulating the handles on the graphs, you can dial up your particular fantasy of a color scheme for the photo—or, more often, repair the color tones in a photo that didn't come out quite right.

- **Noise Reduction** attempts to remove some of the digital "noise" (speckled grain) from a shot.

- **Sharpen** doesn't work miracles—it can't bring an out-of-focus shot into crisp focus—but it applies a modest amount of de-blurring.

At any point during your editing, you can hold down the letter M key (or hold down the ■|□ button at the top left of the screen) to compare your edited masterpiece with the original image. (Professional editors call that kind of comparison "A/B testing"; Apple apparently calls it "M testing.")

So far, you've been reading about the controls on the **Adjust** tab; they're what most people use most of the time. But there are, in fact, two other tabs at the top of the window:

- **Filters.** This is a set of nine fairly random effects that make a picture look washed out, oversaturated, black-and-white, cooler, warmer, and so on. This feature gives you no control over how *much* of this effect you're applying.

- **Crop.** On this tab, you can crop out the edges of the photo, change its angle, or flip it horizontally. In Big Sur, you can even perform these operations on a video, which is a blessing for anyone whose phone has ever incorrectly recorded one sideways.

Drag edges or corners; drag inside the frame to reposition the photo.

Drag to correct a tilted photo.

Choose proportions for the cropping rectangle.

How to crop a photo or video

When you're finished fooling around with your photo in the editing mode, click **Done**, or press the Return key, to return to the regularly scheduled world of Photos.

Preview

This little app of all trades is a universal viewer for graphics, faxes, and PDF documents. Over the years, it has acquired far too many features—this is even the app you're supposed to use to operate a document scanner, if you're one of the eight people who still own one—but it's pretty great at letting you read and mark up PDF documents.

Reading PDF Files

PDF originally stood for "portable document format," and it's used all over the world for preparing and transmitting documents—contracts, instruction manuals, tax forms, and so on.

A PDF document has aspects of a graphic, in that it's essentially frozen in its final form: You can't make changes to it, at least not without specialized software. It looks exactly as the original designer intended, complete with all its fonts, colors, diagrams, columns, and so on. And yet it also has aspects of a word processing document, in that you can search for text in it, fill in blanks, add highlighting or notes, or even drop in your signature. And it can be many pages long.

*Using the **View** menu, you can choose either a **Single Page** or **Two Pages** per screen.*

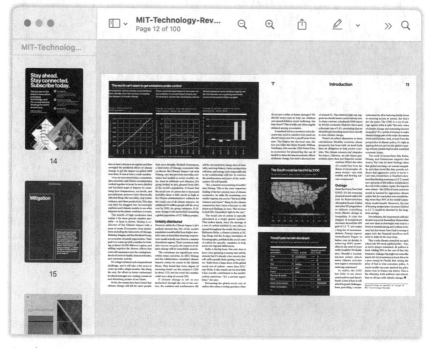

Preview

On the Mac, when you double-click a PDF document, it opens in Preview. Here's how you can get yourself comfortable:

- **Zoom in or out.** Use the View→Zoom commands, or press ⌘-plus and ⌘-minus until the document is at a comfortable size to read.

- **Navigate a long document.** If the document has many pages, you might be delighted to discover that Preview offers a kind of visual table of contents: a scrolling list of thumbnail images of all the pages. If you don't already see this sidebar, choose **View→Thumbnails** to make it appear.

> **TIP:** You can drag these thumbnails up and down to change the page order. You can also select some thumbnails and then press the Delete key to remove the pages from the document. If you have two PDF documents, you can drag the thumbnails from the sidebar of one into the other, thereby moving the pages between the documents. And you can turn a page into a single graphic by dragging it directly out of the Preview sidebar and onto your desktop behind it.

- **Scroll with the space bar.** To move through the document, tap the space bar to scroll one screen at a time. (Add the Shift key to scroll upward again.)

> **TIP:** This is an incredibly useful scrolling shortcut that also works when you're reading emails or web pages.

Marking Up PDF Files

There are all kinds of reasons you might want to add notations of your own to a PDF. Sometimes you've been sent of form to fill in. Sometimes you're looking over a draft of an article, brochure, or ad by somebody who wants your feedback. Often it's a contract, and you're supposed to add your signature to it.

To highlight some text, first select it by dragging through it with your cursor, and then click the ✐ on the toolbar to apply yellow highlighting. (You can use the ⌄ next to it to choose a different color.)

But for all the other sorts of annotations, you must first click the Ⓐ on the toolbar to open the markup toolbar. Here's what you can do now:

- **Add lines, arrows, squares, or circles.** Click the ⬠⌄ to choose a shape; it appears instantly in the middle of your screen. Drag it to move it around, or drag its little blue handle dots to reshape it. You can use the ☰⌄ menu to adjust the thickness of the lines, ☐ ⌄ to choose the color of the lines, or ⬔ ⌄ to fill the inside of the shape with a color.

- **Draw freehand shapes.** If you click the ✐ tool, your cursor becomes a pen for drawing freehand on your document. If you use the ✐ instead, then Preview automatically straightens and perfects whatever you draw. If you drew something like a circle, Preview makes it a perfect circle. If you drew something like a square, Preview sharpens it to a perfect square. Here again, you can adjust the line thickness, line color, and inside color.

- **Add a text box.** When you click ⊡, you get a floating box of text in the middle of your PDF page with the helpful starter text "Text" inside. Simply begin typing to replace it. Use the Aa⌄ menu to choose the typeface and style.

- **Add a sticky note.** Editors, this one's for you. Click ▢ to attach a virtual sticky note to the page; it expands as you type into it. When you're finished editing, click outside the note to make it collapse into a tiny square button that you or somebody else can click to read your observations.

- **Add your signature.** You can teach Preview what your signature looks like in any of three ways, all ingenious: You can write your signature on the laptop trackpad, using your finger as a pen; you can write your signature on a piece of paper and hold it up to the Mac's camera; or, if you have an iPhone, you can write your signature on its screen, and Preview will magically receive it wirelessly.

To begin, click ⌄, choose **Create Signature**, and click the method you prefer: **Trackpad, Camera,** or **iPhone.** The instructions guide you through writing or drawing your signature.

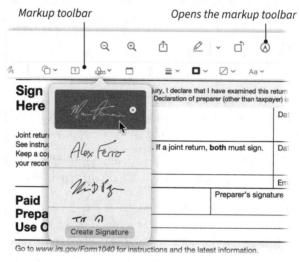

Adding your signature

From now on, you can choose your captured signature from the ⌄ menu. It pops into your PDF document, ready for you to adjust its size and position.

Editing Graphics

Preview is also the app that opens, on most Macs, when you double-click any kind of photo or graphic. (It opens formats including JPEG, TIFF, GIF, Photoshop, and EPS.) You'll quickly discover that it isn't exactly Photoshop, or even Photos. Still, it gives you basic tools like these:

- **Cropping.** Drag diagonally across the piece of the image you want to *keep*. Then choose **Tools→Crop**. Preview eliminates everything outside your rectangle.

- **Color correction.** By choosing **Tools→Adjust Color**, you summon a floating panel filled with sliders that adjust brightness, contrast, sharpness, and so on. The tools work exactly as they do in Photos (page 282)—there just aren't as many of them.

- **Size and resolution.** The **Tools→Adjust Size** command can change the size of a photo (4 × 6, 8 × 10, or whatever) as well as its resolution (the number of pixels per inch). The next time somebody says, "Do you have a smaller version of that photo? It's too big for our website," you'll be ready.

QuickTime Player

If Preview is the Mac's all-purpose graphics and PDF viewer, QuickTime Player is the Mac's all-purpose *video* player. Whenever you double-click a video file, this is the app that opens. Click the ▶, or just tap the space bar, to begin playback.

Whenever your cursor moves in front of the frame, a set of playback controls appears. (To hide them, move your cursor out of the video frame, or just wait a moment.)

You can make the video window bigger by dragging its lower-right corner or any edge—and you can take it full-screen by clicking the green ❷ button at

Picture-in-picture
on your desktop

Send by AirPlay
to an Apple TV

QuickTime Player

upper left. Of course, you can use these techniques in any app. But when the
subject matter is as visual as a movie, enlarging it is especially important.

> **TIP:** You can also adjust the video size by pressing ⌘-plus or ⌘-minus.

Rough Video Editing

iMovie, the video-editing Mac software that has helped creative amateurs
produce millions of great-looking videos over the years, no longer comes
preinstalled on Macs. It's still available, though, and it's still free, and it's easy
enough to download from the App Store.

But if all you want to do is trim the ends (or the middle) of a video, QuickTime
Player can sometimes be just what you need:

- **Trim off the ends.** Choose **Edit→Trim**. Now drag the yellow handles
  inward; everything outside them gets chopped off when you click **Trim**.

> **TIP:** The **View** menu includes a command called **Show Audio Track**. It makes a
> horizontal strip of sound waves appear beneath the scroll bar, which you can use to
> guide your trimming, to avoid chopping off somebody in midsentence.

- **Cut the movie into chunks.** Move the playback cursor to the spot where you want to split the movie. At this point, you can choose **Edit→Split**.

  Now the playback controls show a sort of filmstrip of the whole video, with the two resulting pieces clearly marked. You can double-click either piece to produce trim handles just for it. By adjusting them, you can cut out pieces of the middle of your video. (You can even split the video into, for example, *three* chunks—and delete the middle one by clicking it and then pressing the Delete key.)

  You can also rearrange chunks of your movie using the **Cut** and **Paste** commands in the **Edit** menu, or by dragging them around within the "filmstrip."

  But once your ambitions extend to rearranging clips, using iMovie is a lot simpler and more straightforward.

Recording with QuickTime Player

The name of this app is a little misleading, because QuickTime *Player* is also a great *recorder*. It can use your Mac's camera as a camcorder, it can record the activity on the screen itself, and it can record audio. Here's how:

- **Capture video.** Choose **File→New Movie Recording**. Set up the scene, and then click the red Record button (or press the space bar). To stop recording, click the ⦿ button (or press the space bar again). If your masterpiece is worth saving, choose **File→Save**.

- **Capture audio.** Honestly, Voice Memos is a better app for recording sound. But if you want, you can use QuickTime Player for this purpose by choosing **File→New Audio Recording**. Use the ⌄ menu to specify what microphone you want to use and what quality of sound to record; click the red Record button to begin.

- **Record the screen.** This may be QuickTime Player's best stealth feature. It lets you record what's happening on the screen. The result is a movie file that plays back whatever was happening on the screen—windows, menus, typing, cursor movement, and so on. It's great for creating software tutorials, instructional video podcasts, video clips of computer glitches to send to tech support, and so on.

  Choose **File→New Screen Recording**. At this point, choose what portion of the screen to record—or record the whole screen—exactly as though you had pressed Shift-⌘-5 (see page 389 for the details).

Reminders

Reminders is a to-do app that can remind you about tasks, let you delegate tasks to somebody else, arrange your reminders in multiple lists and folders, and much more.

> **NOTE:** The Reminders version that comes with macOS Big Sur (and macOS Catalina before it) incorporates a new format. It's incompatible with the Reminders app in previous versions of macOS (or iPhone versions before iOS 13). When you first open Reminders, therefore, you may be offered the opportunity to upgrade it from the previous format. You're under no obligation to do so—but if you don't, you'll miss out on some of the better features, like the ability to share Reminders lists with others or apply flags to certain important items (⚑).

Group (folder)   Smart lists

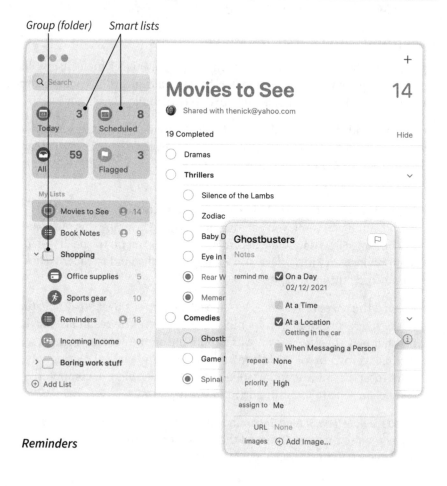

Reminders

Creating Reminders

The most obvious way to enter a to-do item is to choose **File→New Reminder.** A blank line appears at the bottom of the Reminders list, and you can type away.

But there are other ways to create reminders, too:

- **Click below the existing list.** That's actually quicker and easier.

- **Use Siri: "Add" or "Remind me."** Quicker and easier yet: Using Siri (page 371), say, "Remind me to get my oil changed on November 7." "Add yogurt to my Groceries list." "Add 90-inch flat-panel TV to my Birthday Gifts list." You can do this no matter what you're doing on the Mac—Reminders doesn't even have to be open. In fact, you can make these requests when you're nowhere near the Mac, using your iPhone or iPad. Thanks to the miracle of iCloud syncing, each new task shows up identically in Reminders on every device.

- **Use Siri: "Remind me about this."** Here's a supercool option that would probably never occur to you: Suppose you're using one of Apple's apps— like Calendar, Contacts, Books, Mail, Maps, Messages, Notes, Numbers, Pages, Podcasts, or Safari—and something in one of those apps will need your attention later. For example, you're looking at an important text message, email message, Pages document, or web page.

  Without even leaving the app, you can say to Siri, "Remind me about this later." (Or "tomorrow." Or "at 8 p.m." Or "when I get home.")

  Siri creates a new item in Reminders, named for that message or page, and attaches an icon. Later, in Reminders, you can return to the exact message, email, map, note, page, or calendar appointment by opening that icon.

  It's like having a minion with a clipboard scurrying after you wherever you go in your work universe.

- **Select something first.** In many standard Apple apps, including Mail, Notes, Photos, and Maps, you can highlight some text, a photo, or a location and send it directly into Reminders as a new to-do. Right-click or two-finger click the selection, and choose **Share→Reminders** from the shortcut menu. That's yet another cool way to integrate your Mac universe with Reminders.

- **Subscribe to Yahoo or Microsoft Exchange tasks.** Plenty of other, non-Apple online services have their own to-do features—and Reminders can show them to you. To set that up, open **System Preferences→Internet**

Accounts. Click the account you want, like **Exchange** or **Yahoo,** and turn on the **Reminders** switch. (Some of the advanced Big Sur Reminders features may not be available with these lists, but you can always view and check off your to-dos.)

Anatomy of a Reminder

A reminder can be just a few words of text, like "Clean the chimney." But when you click one, five new buttons appear, which you're welcome to exploit:

- **Notes.** Type anything here—an address or phone number, additional thoughts, driving directions.

- **Add Date.** This lets you specify a date and time for this reminder. Later, Reminders can sort your reminders by imminence and urgency.

- **Add Location.** If you specify an address for this item, your Mac (or, more realistically, your phone) will be able to remind you when you get there (or leave there). For example, suppose you've dropped off your shoes for repair at a cobbler's shop. You could set up a reminder called "Pick up shoes!" that pops up the next time you're driving by that location.

  You can type an address here, or you can use one of the choices in the menu: **Current location,** your home or work, or getting in or out of the car (which assumes that your car connects to the phone's Bluetooth).

SHARING AND DELEGATING REMINDERS

When life starts overwhelming you, it's good to know that you can share your task list or even delegate specific items to people.

When you point to one of your reminder lists (not individual items) without clicking, a ⊕ appears. Click it to choose how you would like to invite somebody to access this list—**Mail, Messages, Copy Link,** or **AirDrop.** Click **Share,** address the invitation, and await acceptance.

(Only people with the upgraded Reminders format—iOS 13 or later, macOS Catalina or later—are eligible.)

At this point, both of you have full access to this list. It's a fantastic way to collaborate on lists of chores, gift registries, shopping needs, and so on.

In macOS Big Sur, though, you can take this idea a step further. You can also assign individual tasks to one of your collaborators.

Select the item and then choose **Edit→ Assign [the person's name]**. That person's initials or photo appears next to the item's name, so everybody will know who's responsible.

Delegation: It's a lifesaver.

- **Flag this item.** Click the 🚩. The flag can mean whatever you want, from "super important" to "something to ask the kids to do."

- **Info** (ⓘ) lets you specify even more details about this item. This panel includes the time, place, and flag items already described, but with more detail: For example, the **On a Day** item includes the option to set up automatically repeating reminders.

  When Messaging a Person is incredibly cool. Click **Add Contact** and choose somebody from your contacts list. The next time you're in Messages chatting with this person, this reminder will pop up automatically, saying "Chris still has my bike" (or whatever you wanted to be reminded about).

  Here, too, you can set a **priority** level for this item, enter a web address (**URL**), or attach a picture.

Organizing Your Reminders

You can tweak your reminders in a thousand different ways. For example:

- **Rearrange your items.** Drag reminders up and down within a list. (Use any blank spot on the line as a handle.)

- **Create new lists.** Every time you choose **File→New List,** you create a new, independent list. One might be a shopping list. Another might be movies you want to see or books you want to read. Maybe one of them is a list of things you have to do to prepare for a party or a meeting.

- **Create a subtask.** If you choose **Edit→Indent Reminder,** you nest the selected item *beneath* the one above it, creating a subtask. The main reminder might be "Prepare car for the drive," and the subtasks might be "Change oil," "Repair flat," and "Replace engine."

> **TIP:** You can achieve the same effect by just dragging one task beneath another, using any blank part of the line as a handle.

- **Group your lists.** If you choose **File→New Group,** you get a little folder in your list of lists. You can drag individual lists onto this folder, thereby grouping them. For example, you might have a group called **House Stuff,** and the to-do lists within are **Things to Clean, Things to Fix, Things to Upgrade,** and **Things to Pay For.**

As you would hope, all this organizational effort magically reproduces itself in Reminders on all your other Apple gadgets—phone, tablet, and other Macs.

Marking To-Do Items as Done

Obviously, you check off a to-do item by clicking the empty circle next to its name. A satisfying little dot appears, and the item vanishes. But, depending on your personality type, you may prefer that checked-off items don't disappear so you can look over your accomplishments with pride. In that case, choose **View→Show Completed**.

You can also just delete items without doing them. Click a blank area of its line and then press the Delete key.

Stickies

As though Notes and Reminders don't already give you enough tools for jotting down random thoughts, this virtual notepad is at your disposal. Create

Collapsed note (double-click title bar)

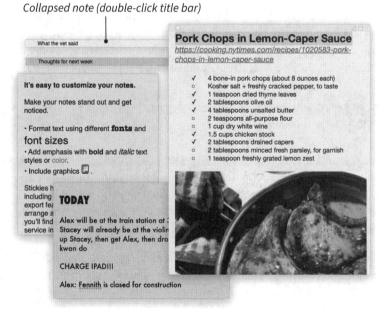

Stickies

a new floating note by choosing **File→New Note**. Close one—thereby deleting it—by clicking it and then clicking the tiny square at top left. Change the font and color using the menus.

Fill a note by typing, pasting, dragging text or graphics from other apps, or using the **Make New Sticky Note** command. You'll find it in the **[Application Name]→Services** submenu of most apps. For example, in an email, highlight some text and then choose **Mail→Services→Make New Sticky Note**.

> **TIP:** If you double-click the narrow title bar of the note, you collapse it into a skinny floating bar—handy when you want to get it out of your way. You still see the first line of text, though, so you can remember what's in it.

Stocks

As you could probably guess, this app lets you track your favorite stocks (or your least favorites).

Zoom in or out of the
graph's time scale.

Drag across the graph to see
gains or losses in that period.

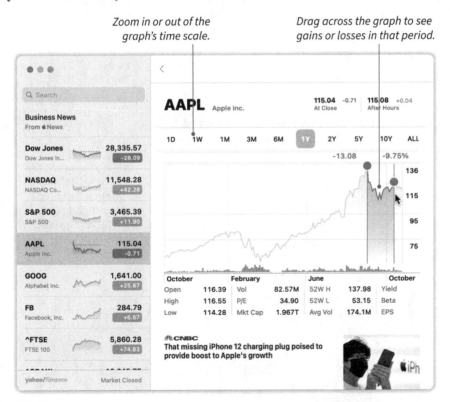

Stocks

Click the name of a stock in the sidebar to view a beautiful graph of its price over time; it's red if the stock went down today, green if it went up. Above the graph, buttons like **3M** (three months) and **2Y** (two years) adjust the timescale.

As you move your cursor (without clicking) over the graph, you see the price indicated for that time. And if you *drag* across the graph, you see the stock's performance during the time range you've highlighted.

To remove one of the stocks from the list, click it and then press the Delete key; confirm by clicking **Remove**.

To add a new one, use the search box above the list. Type a company name or stock symbol. Click the company's name in the search results, and then click **Add to Watchlist**.

TextEdit

What would your computer be without a basic word processing program? This is it. You can type, check your spelling, and incorporate graphics, tables, and formatting. You can even open and create Microsoft Word documents— no charge.

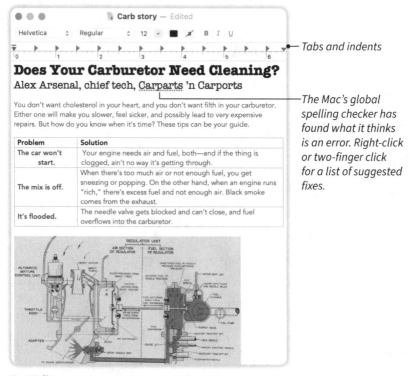

Tabs and indents

The Mac's global spelling checker has found what it thinks is an error. Right-click or two-finger click for a list of suggested fixes.

TextEdit

It's easy enough to get going: Choose **File→New**. On the new blank page, start typing. Or paste something in from another app. The toolbar offers a full range of formatting options, like bold, italic, fonts, sizes, paragraph alignment, bulleted or numbered lists, line spacing, and colors.

> **NOTE:** Be aware that TextEdit can also serve as a *plain text* editor, a program that doesn't permit any formatting. If you don't see the formatting toolbar and can't seem to change your formatting or typeface, it's probably because you're editing a plain-text document. Use the **Format→Make Plain Text** or **Format→Make Rich Text** commands to convert your document between the plain and formattable types.

The **Edit** menu can take care of all your searching and spelling/grammar-checking needs; the **Format** menu lets you insert a tidy **Table** with any number of rows and columns, or a neatly numbered or bulleted **List**.

When your masterpiece is complete and you choose **File→Save**, you'll see that the **File Format** pop-up menu lets you save it as, among other formats, a Microsoft Word (.docx) document, which is the most common word processing format in the galaxy.

Voice Memos

Most Mac models come with a built-in microphone—as well as this little app, which is your recording booth. It's great for recording performances, conversations, lectures, and cute things your kids say. There's an identical app on the iPhone and iPad (and even Apple Watch), which synchronizes those recordings with your Mac, and vice versa.

To make a recording, click the big red Record button. You can click **II** and **Resume** as often as you like; click **Done** when the recording session is over.

The new audio masterpiece appears in the sidebar with a generic name; click it to rename it. Now you can go to town:

- **Play it.** Click a recording and then click ▶ (or press the space bar) to listen to it.

- **Jump around.** Click in the scrubber "map" to skip past the boring parts in the recording. Or click ⑮ or ⑮ to jump forward or back by 15 seconds.

- **Share the audio.** The ⬆ button, as always, can send your recording to someone else by AirDrop, email, or Messages—or you can use it to drop the audio file into one of your Notes pages.

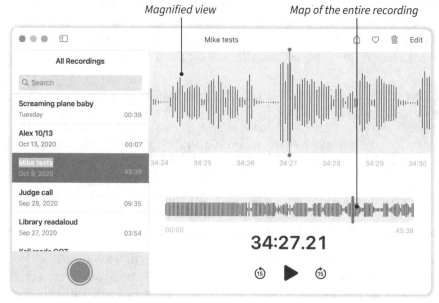

Magnified view   Map of the entire recording

Voice Memos

- **Delete a recording.** Click its name in the list and then press Delete. Voice Memos has its own Trash (called **Recently Deleted**); you have 30 days to rescue anything inside.

Editing Recordings

If you click a recording and then **Edit** (top right), you open a special Edit screen. Here, the audio "map" features a blue handle that you can use to position the blue playhead bar.

- **Record over a part.** When the blue playhead line is parked at the beginning or middle of the audio, you can click **Replace** to record over your original recording. That way, you can re-record only the part where you messed up. Hit ❙❙ to stop.

- **Add to the end.** Position the blue playhead line at the end of the current sound waves and then hit **Resume**. Click ❙❙ to stop.

- **Trim the beginning or end.** Click ⛶ at top right. Now the little map of the recording sprouts yellow crop handles. Drag inward to isolate the part you want to keep. Use the magnified area at the top to fine-tune the trim points. When you've neatly bracketed the worthy portion, click **Trim**. If all is well, hit **Save**; if you messed up, choose **Edit→Undo**.

- **Snip out the middle.** To cut a bit of audio out of the middle—that car horn that interrupted your wedding vows, say—isolate the bad spot using the trim handles as already described. This time, though, click **Delete**. Only the outer chunks remain.

> **TIP:** If there's background noise or room echo, try the new **Edit→Enhance Audio** command. Apple says it uses artificial intelligence to clean up static and reverberation. *You* may say the result is barely noticeable or even makes things worse; in that case, you can always use the same command again to turn it off.

Editing audio

The Utilities Folder

That's still not the end of the software cornucopia that came with your Mac. Inside the Applications folder is another one called Utilities, filled with advanced diagnostic and troubleshooting apps that most people never touch.

There are, however, two apps in here that you may occasionally be directed to consult:

- **Activity Monitor** shows a list of all the software running on your Mac right now and how much of your processing power each one is using. What might startle you is how *many* programs are running—because behind the scenes, the Mac runs a long list of software bits with technical names and no icons or windows. These constitute the background activity of your Mac: making Time Machine backups, checking for email, listening for Siri, and so on.

Across the top: tabs that let you view the running apps according to how much **Memory, Energy, Disk** space, or **Network** capacity they're using.

At some point in your Mac life, some technically proficient person might ask you to open Activity Monitor and report on one of these readouts.

Quit or force-quit this app

Process Name	% CPU ⌄	CPU Time	Threads	Idle Wake Ups	% GPU	GPU Time
Activity Monitor	13.1	13.79	7	3	0.0	0.00
Messages	6.1	29:54.02	9	66	0.0	0.01
ForteDaemon	3.8	1:07:26.94	7	57	0.0	0.00
CEPHtmlEngine	2.5	24.03	19	61	0.0	0.00
Voice Memos	1.6	22.57	8	79	0.0	0.00
Mail	1.3	31:18.61	16	1	0.0	0.11
AccessibilityVisuals...	1.0	1:09.27	4	0	0.0	0.00
AXVisualSupportAg...	0.9	4:23.08	5	10	0.0	0.00

Activity Monitor — My Processes — CPU Memory Energy Disk Network

		CPU LOAD		
System:	4.25%		Threads:	3,269
User:	8.19%		Processes:	753
Idle:	87.56%			

Activity Monitor

- **Terminal.** You might think of macOS as a beautiful operating system, constructed of gorgeous shapes, tasteful colors, drop shadows, and elegantly designed icons. But all that is a veneer, a visual fiction, presented for you, the human being.

  Behind the scenes, what's really running your Mac is Unix, an industrial-strength, decades-old operating system that drives the nation's banks and universities. It definitely does not have tasteful colors and drop shadows—in fact, it has no graphics, menus, or windows at all. What it does have is a lot of power. You can do things in Unix that you can't do at all with the mouse.

  Your window into the Unix world is Terminal. It's of no value to you unless you know exactly what Unix commands to type—but here and there, when troubleshooting or when reading articles online, you'll be given instructions to do just that. And now that you've located the Terminal app, you'll know how to begin.

PART FOUR

Beyond the Basics

CHAPTER THIRTEEN

Accounts and Security

N o big tech company on earth advertises its efforts to keep your data private and secure the way Apple does. Maybe that's because Apple's heart is in the right place, sure—but maybe it's also because Apple has no *reason* to collect your data. Companies like Facebook and Google harvest your data and track your actions because they're in the business of selling targeted ads; Apple is in the business of selling machines and services.

At any rate, privacy and security begin at home—and on the Mac, they begin with *user accounts*.

Introducing Accounts

Like any self-respecting computer these days, every Mac is designed to be shared.

You can log in with your name and password, and boom: There are all your files and folders, your desktop wallpaper and other settings, your web bookmarks, your email accounts, your calendar and notes and reminders, even the fonts you like to use.

But later, when your sister logs in with *her* name and password, she sees a completely different world—her own files and settings. Your stuff is completely invisible to her.

Various people can use the same Mac at different times without ever encountering (or messing up) anyone else's stuff. That's a huge benefit for classrooms or workplaces.

Now, if you're the *only* person who uses your Mac, you probably don't care about all that account-switching stuff.

But even for you, the name-and-password business is useful, because it protects your Mac from snooping by ne'er-do-wells who come poking around when you're away.

Logging In

Once you have an account, logging in is easy. On the login screen, click your account name, enter your password, and press Return.

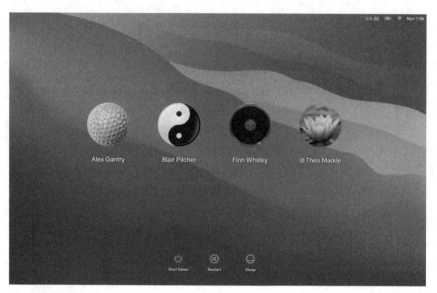

The login screen

If you enter the wrong password, the whole password box "shakes its head"—shudders horizontally as though to say, "Nope, try again!" After another incorrect guess or two, the Mac may offer your password *hint* or the option to reset your password.

> **TIP:** If you have a Touch ID laptop, you can combine both steps. When you touch the fingerprint sensor, the Mac figures out who you are *and* shoots past the password screen.

Creating an Account

The first time you ever powered up your Mac, the setup software asked you to make up a name and password for yourself. That's when you created your own user account. To this day, everything in your home folder belongs to your account, and nobody else can see it.

There may come a time, however, when you want to create another account. Maybe somebody new has entered your life (or classroom, or office), and you'd like to offer them their own account on the Mac. Here are the steps:

1. **Open System Preferences→Users & Groups.**

   Here's your Mac's headquarters for account administration.

2. **Unlock the Users & Groups panel.**

   There's immense power in this control panel: If you had blackness in your heart, for example, you could delete somebody else's account, and every file, photo, and folder they own. That's why Apple demands that you prove you're somebody with authority. You're required to click the padlock at lower left and fill in your account password.

3. **Click + beneath the accounts list.**

   An important little panel appears. Here, using the New Account pop-up menu, you must choose among four options:

   What kind of account is this? An **administrator** account is the kind that, presumably, you have. You're the Mac's owner. You're in charge. In a family, you're the parent. In a classroom, you're the teacher. Only you are

Click to create an account.

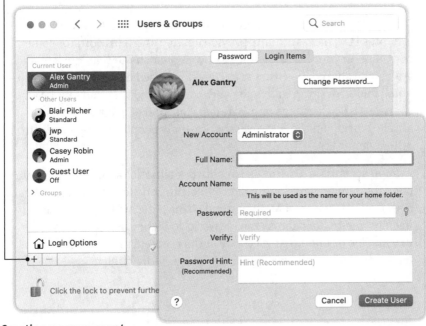

Creating a new account

allowed to install new apps, add or delete accounts, or change important settings.

It's perfectly legal to bless another account holder with administrator powers, as long as you're aware that you're giving them great power and great responsibility.

To everybody else—your kids, your students, your employees—you should give **standard** accounts. These accounts have a different set of powers. These people are allowed to do anything they want to their *own* files, folders, and settings. But if they try to make any changes to the Mac's settings that affect everybody, like changing the time zone or deleting an app, they'll be blocked at the gate. They'll encounter the 🔒, and they'll be forced to ask you, or another administrator, to come over and approve what they're trying to do. You'll have to enter your name and password to grant that approval.

MacOS seeks an administrator's password—or fingerprint.

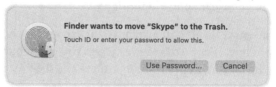

Administrator approval

Sharing only accounts are useful only if your Mac is on a network (Chapter 14). These accounts provide a name and password for some-body else on your home or office network, just so they can duck in and grab some shared files from you. They don't need a full-blown account—a home folder, desktop wallpaper, and all that—and they can't make any changes to Mac settings or programs. This kind of account exists *only* for file sharing on the network; sharing-only people can log in *only* from other computers.

A **group** account is also for use in network situations. You might create one group for all your kids (called Offspring), one for your most critical staff (called Cherished Minions), and so on. Later, when you're shar-ing files and folders across the network, you'll be able to grant access to some file or folder to everybody in the group in one step, instead of having to fiddle with each member's access individually.

When you type a name for your group and click **Create Group**, the Mac reveals a list of all existing account holders on this Mac. Turn on the checkboxes to indicate who belongs to this group.

4. **Make up a name and password.**

Full Name is usually the first and last name (*Chris Robertson*, say); the **Account Name** is usually something shorter (*Chris*—or, in a classroom, maybe *ChrisR*).

Password is the password to log into the Mac, of course. But macOS will also require you to type it before you make certain changes, like deleting important Mac folders and, yes, making changes to the Mac's accounts. You have to type the password twice here, to guard against typos. (And you see only dots as you type, to guard against somebody watching over your shoulder.)

If you work at the CIA, your password should be long and complex. If it's just you and your life partner, or you and a co-worker you trust, the password can be simpler. It can be just a letter or two, for all the Mac cares.

When you're trying to log in, and you enter the wrong password a couple of times, the Mac will display the **Password Hint** to remind you. It can be whatever memory-jogger will help you in times of forgetfulness.

LOGGING IN AUTOMATICALLY

The login screen appears whenever you turn on the Mac or log out of your account. Here's where you choose your account and supply your password.

But in **System Preferences→Users & Groups→Login Options**, you can adjust the trade-off between convenience and security. If you're the only person who ever uses your Mac, why should you have to log in every time you sit down?

Believe it or not, the automatic login option lets you *skip* the login screen. When you turn on the Mac, it logs you in automatically, and you whip straight to the desktop, ready to roll.

Unfortunately, automatic login is incompatible with Touch ID and Apple Pay.

If you turn on **Automatic login**, you're stripping the Mac of an important security barrier, so Apple turns *off* the options to log in with your fingerprint (page 76) and use Apple Pay (page 77). You'll have to enter your iCloud password every time you buy anything from one of Apple's online stores—books, apps, movies, and so on.

Still interested? Choose your name from the **Automatic login** pop-up menu, enter your password, and click **OK**. From now on, the Mac logs you in automatically when you turn it on.

The login screen still appears when somebody logs out manually by choosing →**Log Out**, for example. ★

5. **Click Create User.**

   Now, back at the Users & Groups pane, the new account name appears in the list of accounts.

6. **Set the options for this account.**

   If you turn on **Allow user to reset password using Apple ID**, then if this person ever makes three incorrect attempts to enter her password, the login screen will offer her a chance to recover it by entering her Apple ID on a website.

   Allow user to administer this computer turns a standard account into an administrator account.

7. **Choose an account picture.**

   Every account has an associated picture. It appears on the login screen, for example, and it becomes your icon in Messages, Mail, and Contacts.

Take a photo now.   *Grab a picture from Photos.*

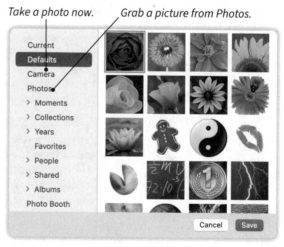

Choosing an account photo

System Preferences starts you out with a lovely little nature photo, but you can substitute a photo of yourself—or of anything else.

To choose a photo, click the starter image. In the resulting box, you can choose from **Defaults** (generic stock photos of flowers, sporting gear,

---

THE GUEST USER ACCOUNT

Suppose a buddy is at your house, waiting for her ride home. You want to offer her your Mac to use while she waits, but you don't want to worry that she's going to root through your stuff. Creating a whole new account for her is probably overkill. What's your play?

You'd let your guest use the guest user account, of course. That's a *very* temporary account that leaves all the other accounts private and protected—and when your buddy logs out, no shred of her activity is left behind. Any changes she makes are erased, any files she made are vaporized. The guest user account is left pure and new for the next guest to come along.

There's only one guest user account, and it's listed in the main Users list. It's

dormant until you click it and turn on **Allow guests to log in to this computer**.

If your guest is a kid (or if you have high morals), you can also turn on **Limit Adult Websites**.

If you're OK with letting the guest exchange files with your Mac from across the network (Chapter 14), turn on **Allow guest users to connect to shared folders**.

And that's it. Now, on the login screen (page 316), your friend can choose **Guest User** and have full access to all the apps on your Mac, surf the web, check her email, or whatever.

But when she signs out, any evidence that she used your Mac at all is wiped away for good. ✦

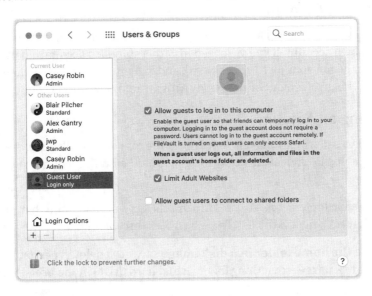

wildlife), **Camera** (take a new photo right now, using the Mac's camera), or **Photos** (your entire collection, as it appears in the Photos app).

In each case, you get the chance to adjust the photo's size within the round frame (by dragging the slider) and its position (by dragging the photo). Click **Save** when it all looks good.

8. **Set up your login items.**

This feature can be a great time-saver: The Mac can auto-open certain apps or documents automatically when you log in. For some people, that might be Safari and Mail. For others, it might be Reminders and the book manuscript they've been working on.

> **NOTE:** Each account holder must set up his own login items when he's logged in. The administrator can't do it.

To choose your login items, click your account's name and then click **Login Items**. Click + to choose the first app, document, disk, or folder you want to auto-open each time you log in. (Alternatively, you can drag any icon from the desktop right into this list.)

If you turn on the **Hide** checkbox, the app shyly steps to the background after auto-opening—a useful feature for utility apps that do their work in the background, like Dropbox, typing expanders, or keyboard-macro programs.

Editing an Account

If you have an administrator account, you can make any changes you like to your own account. If you've logged in with a standard account, all you can change are your picture, password, and login items. Any other changes require the involvement of somebody with an administrator account.

Deleting an Account

It's always sad when someone leaves your life for good. (Well, "always" might be stretching it.) At least it's easy to delete that person's account from your Mac.

Open **System Preferences→Users & Groups**, click the account, and then click the − button beneath the list. The Mac offers three options:

- **Save the home folder in a disk image** saves the deleted person's stuff in a disk image (something like a virtual hard drive) in the **Users→Deleted**

Users folder. Later, you can open this .dmg file by double-clicking, and look through the stuff in this "disk."

- **Don't change the home folder** leaves the person's home folder, containing all her stuff, right where it is. But the person's account, as listed in the Login list and in the Users & Groups panel, is gone.

- **Delete the home folder.** This option nukes the account and all its files and settings forever. (Also turn on **Erase home folder securely** if you want the Mac to scrub over the spot on your drive where that home folder once was, so no recovery software could ever get it back.)

> **TIP:** If you have a laptop, take this opportunity to add an "If found" message to the login screen. It might include your phone number, your email address, or an offer to provide a reward. You can set this up in **System Preferences→Security & Privacy→General**. Click the 🔒, authenticate yourself, and then turn on **Show a message when the screen is locked**. Click **Set Lock Message**, and then type your message.
>
> If the person who finds your laptop has any decency at all, you now have a prayer of getting it back.

All About Logging Out

When you're finished with your work session, you can log out. The Mac seals up your account against marauding snoopers. The login screen returns, ready for its next victim. Here's how to log out:

Logging out

- **Choose ◆→Log Out.** In the confirmation box, click **Log Out** or press Return.

  Or, if you hold down the Option key as you choose ◆→Log Out, the confirmation box doesn't appear. You save a step.

- **Set up auto-lock.** You can also tell the Mac to lock up your account automatically after a period of inactivity. That's a nice safety net if you tend to wander away from your desk and forget to come back.

  Open **System Preferences→Security & Privacy→General.** Click the 🔒, authenticate yourself, and then click **Advanced.** Turn on **Log out after [60] minutes of inactivity.** Set the time to whatever interval of inactivity you like; 5 or 10 minutes works well.

After that, the Mac signs you out automatically, and the next person who comes along will find only the login screen.

> **TIP:** If security is all you're worried about, you'll save time by setting a password-protected screen saver instead of making the Mac log you out all the way. See page 62.

Sharing Across Accounts

OK, so the Mac keeps each family member's stuff in a separate, password-protected silo. Very nice. Nobody can look at anybody else's stuff. Fine.

But what if you *want* other family members to be able to see a file or a folder? Whatever happened to collaboration? That's what the Shared folder is for. To find it, in the Finder, choose **Go→Computer**; open **Macintosh HD→Users**.

Anything anyone puts into the Shared folder is available to every account holder, either on the Mac or over the network. Well, "available" is pushing it. In fact, people can look at or copy these files and folders, but they can't delete them or make changes. Still, it's a great sort of public square for commingling of files across accounts.

> **TIP:** Actually, there's another option. Inside your home folder, inside your Public folder, a little folder called Drop Box awaits (not to be confused with the commercial Dropbox service).
>
> Any files or folders other people drop into your Drop Box become available to you, but nobody else. It's a wormhole between Mac accounts.

Fast User Switching

Suppose you're anxiously awaiting word on whether you've gotten that job, that school admission, or that love interest you're obsessed with. You want to check your email, but your kid has been sitting at the Mac for hours.

For this situation, Apple has invented a way for you to log into your account, check something, and log out again—without your kid having to log out and close everything down. In fact, his account is

Currently logged in

Fast user switching menu

still alive and running; if his apps were downloading something or processing something, they keep right on downloading or processing even while you're working in *your* account.

Just click the ⊙ icon and, from the pop-up menu, choose your account name. Log in as usual. Go about your business. When you're finished, log out.

Your anxious kid, standing by through all this, can choose his own icon and sign in again. He'll discover that his whole work environment is still open, exactly as he left it. He was never logged out—he was only fast user switched.

> **TIP:** In **System Preferences→Users & Groups→Login Options**, the **Show fast user switching menu as checkbox** makes the switching menu appear or disappear. The **Show fast user switching menu as** pop-up menu controls whether the menu shows up as a ⊙ icon or the account name.

Security, Privacy, and You

Let's face it: *Security* usually means *hassle*. Logging in and out means waiting. Passwords are a pain. Two-factor authentication (page 151) takes an extra step.

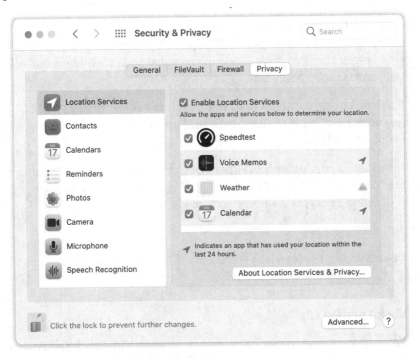

The Privacy pane of System Preferences

Apple has done its darnedest to build in state-of-the-art privacy and security while keeping the inconvenience to a minimum. For example, the Mac can generate (and memorize) complex passwords for you (page 328); it can warn you when a data breach has exposed one of your passwords to hackers (page 199); and it's constantly asking for permission to let a new app use, for example, your camera or your microphone.

But those are only the beginning.

Gatekeeper: Malware Defenses

The bad guys of the world have spent more of their efforts trying to infiltrate Windows machines than Macs. One reason: There are many more Windows PCs in the world than Macs. The other reason: The Mac is harder to crack.

Part of that toughness comes from Gatekeeper, the Mac's app-watching overlord. Its job is to prevent any known malware app from opening—or even any app that comes from a source Apple hasn't vetted.

To learn more about Gatekeeper—and to find out how to override its recommendations when you're confident about the innocence of a new app—see page 352.

FileVault

FileVault is a feature that encrypts your entire hard drive and everything on it. Even if some spy from SPECTRE slips into your home, pries apart your Mac, and *steals* the hard drive, she won't be able to get anything useful off of it. Your data is really, really safe.

On one hand, worrying about hard-drive theft is a pretty paltry piece of paranoia for most people. Who's really that interested in your shopping lists and photo collections?

On the other hand, this layer of supersecurity imposes very little inconvenience to you. You use your Mac as usual. MacOS decrypts each file as you open it and re-encrypts it when you close it, instantaneously.

To turn FileVault on, proceed like this:

1. **Open System Preferences→Security & Privacy→FileVault.**

   Click the 🔒 and then authenticate yourself.

2. **Click Turn On FileVault.**

   For each existing account on your Mac, you now must click **Enable User** and supply that account's password. That's to confirm that no forgotten passwords are kicking around—because without the password, each account is about to be scrambled and lost *forever*.

3. **Click Continue, and set up your back door.**

Encrypting all your files with military-grade encryption is scary business. If you lose your password, all your files are *gone for good*.

That's why, at this point, macOS invites you to create a sort of back door—a master password that can get past the FileVault encryption even if someone forgets their password.

Your options are **Allow my iCloud account to unlock my disk**—the simplest and easiest option—and **Create a recovery key and do not use my iCloud account**, which supplies you with a long, complicated override code like RK94-YFX9-XN3K-LVT5-PL9N-LT53. You're supposed to write this down and not lose it.

4. **Click Restart.**

When you log in next, the Mac begins encrypting everything on your Mac. The process can take many hours, but you're free to keep working on your Mac while it's happening.

From now on, automatic login isn't available, and you have to enter your password every time you wake the Mac or exit the screen saver. But at least you'll know that your entire Mac's contents are encrypted well beyond the capacity of the evildoers of the world to unscramble.

THE FIREWALL: YOUR SOFTWARE BARRICADE

Used to be that hackers could send out electronic feelers into people's computers through their cable modem or DSL box and—if the machine's security wasn't great—even take control. The invention of the barricade known as a firewall put a stop to that sort of infiltration.

These days, most homes' internet setup includes a *router*—a box that distributes the incoming internet signal to multiple computers, usually through a Wi-Fi base station. And every router already has a built-in firewall, meaning all your machines are protected.

If, by some freak of setup, you have a broadband connection but you *don't* have a router, then you can use the Mac's built-in firewall feature instead. Open **System Preferences→Security & Privacy→Firewall**, click the 🔒, authenticate yourself, and then click **Turn On Firewall**.

If you click **Firewall Options**, and you've got some experience with network security, you can now fine-tune the Mac's firewall, allowing or denying internet connections individually for *each app* you use. ★

The Keychain

As you know from Chapter 9, Safari saves you untold migraines by memorizing website passwords for you and autofilling them when you try to visit those sites. But macOS can memorize all kinds of other passwords for you, too: for shared folders on the network, for Wi-Fi networks, for your email accounts, for encrypted disk images in Disk Utility, for FTP sites, and so on.

What's cool is that the simple act of logging into your account simultaneously unlocks all those *other* passwords. Your account password is the master lock.

Ever wonder where this seething mass of memorized passwords winds up? It's in an app called Keychain Access, which resides in your **Applications→ Utilities** folder. Click **Passwords** to see the list of all memorized passwords.

You can search, sort, slice, and dice this list. You can also view any of these passwords; double-click its name, hit **Show password**, and log in with an administrator's account name and password.

> **TIP:** Keychain Access is also a good, very secure place to record credit card digits, ATM codes, or your innermost longings and desires. If you choose **File→New Secure Note**, you get a tidy, password-protected page that you can fill with private things.
>
> If you're especially alert, you may be objecting that password-protected Notes (page 277) do exactly the same thing. True! But someone sitting next to you can *see* that a note is protected and might wonder what you have to be so cagey about.
>
> On the other hand, nobody would ever think to look for secrets in Keychain Access. It's a lot harder for someone to guess your secret if they're not even aware you have one.

Apps' Access to Your Mac

To Apple, security is protecting your Mac. Privacy is protecting your data.

In **System Preferences→Security & Privacy→Privacy,** you can see a huge list of Mac features and software that apps may seek to access. A weather app might need your location. A video chat app needs your camera and microphone. A time-management app might want your Calendar data.

You may get a lot of interruptions from apps asking for access to these components. It's all designed to keep you aware of which apps are using what—and, by alerting you, to prevent a malevolent app from sneaking access to your private data.

You can visit the Privacy pane to see, in each category, the apps to which you've granted permission—and, if you're nervous about one, turn off its checkbox, thereby shutting down its access.

CHAPTER FOURTEEN

Networks and File Sharing

J ohn Donne hit the nail on the head when he wrote, in 1624: "No Mac is an island, entire of itself; every Mac is a piece of the continent, a part of the main." Something like that.

Clearly, he was referring to the importance of *networking*: connecting computers so they can communicate, either using wires or wirelessly using Wi-Fi. Networked computers can share files, even if some are Macs and some are Windows PCs. They can display a mutual calendar or play tunes from a mutual music collection. They can play games against each other. They can share equipment, like printers or internet connections. They can share their screens, so somebody who's technologically competent can help out a co-worker down the hall.

You've probably used at least one network already—the really big one known as the internet. This chapter starts by describing how to connect to it.

Joining the Network

The world's stores, homes, and offices offer networks in one of two flavors: wireless or wireful.

Wi-Fi Hotspots

Wi-Fi networks use radio waves to connect computers, printers, and other network gear. If you've ever connected to a hotspot in a coffee shop, airport, or friend's house, you're already familiar. (A typical Wi-Fi base station—the piece of gear that's the source of a Wi-Fi signal—has about 300 feet of range.)

Now, if you're in range of a hotspot you've used before—your home office, for example, or a hotel you've stayed in—the Mac auto-joins it. You get no notification or password request; just by walking into the room, you get online.

If you're in a new place, though, your job is a little harder. To see what hotspots might be available, click 🛜 on your menu bar. Here, in a tidy menu, is a list of all the hotspots whose signal the Mac has picked up.

> **TIP:** If you don't see the 🛜 menu, you can also use the 🎛 to choose a hotspot. Click **Wi-Fi** to reveal exactly the same options the 🛜 menu would show you.

Wi-Fi menu

The 🛜 menu may be divided into several sections:

- **Personal Hotspots.** The Personal Hotspot feature, also known as tethering, lets your iPhone (or cellular iPad) serve as an internet antenna for your Mac. Because the phone can get online anywhere there's a cell signal, Personal Hotspot means your laptop can get online in far more places and situations than it could over Wi-Fi alone. Details are on page 79.

  In the 🛜 menu, each nearby phone's name appears along with its signal-strength and battery indicators.

- **Preferred Network** is one you've used before, whose password the Mac already knows.

- **Other Networks** refers to nearby Wi-Fi hotspots you've never used before.

In each case, a 🔒 indicates a hotspot that requires a password, and a blue icon indicates the Wi-Fi network you're already on.

If you've just wandered into the range of a new Wi-Fi hotspot, the Mac generally displays a message with its name. Double-click it to connect.

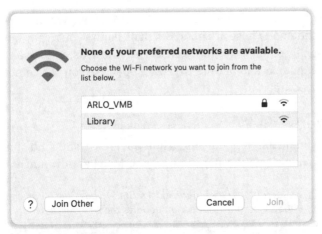

New hotspots found

WI-FI TROUBLESHOOTING BASICS

There are all kinds of reasons you may seem to be on a Wi-Fi hotspot but you can't actually get onto the internet. For the sake of your sanity, here's a guide to all the things that can go wrong.

The first one is obvious: Most Wi-Fi hotspots these days don't let you connect without a password, which you'll have to get from the homeowner, office staff, waiter, or whoever pays the bills. You'll know this is the problem because there's a 🔒 next to the hotspot's name in your 📶 menu.

Remember, too, that the number of signal-strength bars (📶) on your menu bar doesn't indicate how strong your *internet* signal is. It shows only how strong the *Wi-Fi transmitter's* signal is,

wherever it sits in the building. That transmitter (base station) might not even be connected to the internet at all! Even if the internet's out in your neighborhood, you may still see a full three-bar signal on your 📶 menu. It's just saying you're well connected to the base station; it says nothing about the internet connection beyond that.

In some situations—data-sensitive offices, for example—hotspots may have been set up to connect only to *specific* computers and yours isn't one of them.

Finally, you might be connected to a commercial hotspot but you haven't yet signed in using its Welcome web page. Pull out your credit card and get going. ✦

Sometimes, as in a hotel or airport, you'll find yourself within range of a *commercial* hotspot. These are hotspots that require you to pay money, look at an ad, sign in, or all of the above. In these situations, as soon as you choose the network's name, a miniature web page opens so you can complete the sign-up procedure.

Ethernet Networks

Wi-Fi is great and all. Hopping onto a hotspot is easy and kind of magical. But plugging in a physical Ethernet cable to your Mac has advantages.

Mac laptops lack Ethernet jacks. But $15 buys you a USB or USB-C adapter that accommodates the network cable.

Ethernet adapter

A cable connection is usually faster than Wi-Fi, for example, because the internet signal isn't leaking all over the building. It's also far more secure; bad guys have been known to "sniff" Wi-Fi transmissions, but sniffing your cables won't get them anything (unless they're cocker spaniels).

Cables are more reliable, too. With Ethernet cables, you never encounter those weird situations (as with Wi-Fi) where you *seem* to be online, but you can't pull up a website.

Three Ways to Share Files

One of the most popular features of having a network is file sharing: the ability to pass files and folders back and forth or to put files and folders into one central place for communal editing. Macs and even Windows PCs can join in the collaboration.

Over the years, Apple has come up with three different ways to share files from the Mac. They're all still around, each with its own set of advantages:

- **AirDrop** lets you shoot files or folders wirelessly between nearby Apple gadgets: Macs, iPhones, iPads. This is incredibly easy and convenient—no passwords, no setup, no hassle—but limited to Apple machines. And you're actually handing copies of your files off to other people, rather than leaving them on your Mac so multiple people can see them simultaneously.

- **The Public folder** is a special folder. Anything you put in here, other people on the network can see and edit—no password or fuss. (You do have to *put* the shared files here, which may mess up your folder-organizing structure.)

- **The "any folder" method.** This system is far more complicated but also far more flexible. It lets you make *any* file, folder, or disk available for inspection by other people on the network, without having to move it into a special folder. You also get elaborate control over who is allowed to do what to your files. Can Alex see the files? Can Chris edit them? You can make these decisions on a colleague-by-colleague basis.

The following pages describe these three methods in turn.

AirDrop

Somebody at Apple deserves a raise for this feature. AirDrop sends files to other people's Macs, iPhones, or iPads up to 30 feet away, instantly and wirelessly. It's a breakthrough in speed and simplicity. There's nothing to set up, no passwords involved, and no security risk—both sender and receiver are in total control. You don't need to be in a Wi-Fi hotspot. In fact, you don't even need an internet connection; AirDrop works when you're out at sea, in a cornfield, or in a hot-air balloon.

AirDrop is fantastic for firing a file or folder from the Finder off to a friend. But it's also built into most of Apple's apps, so you can send a note from Notes, a photo from Photos, a map from Maps, and so on.

Here's how to send stuff wirelessly using AirDrop:

1. **Open whatever you want to send.**

   If the idea is to send a file or folder from the Finder, click to select its icon. If you're trying to send a photo or video, website, map location, or note, get it on the screen in front of you.

2. **From the Share shortcut menu or the 🗘 menu, choose AirDrop.**

   That shortcut menu appears when you right-click or two-finger click a file or folder icon in the Finder. Choose **Share→AirDrop**.

The Share→AirDrop command pops up in apps all over the Mac.

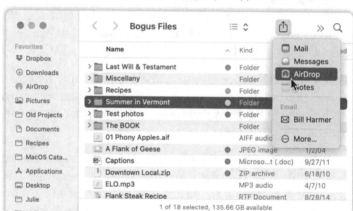

AirDrop in the Share menu

But the 🗘 icon is hiding all over the Mac, too. You can find it on the toolbar of every Finder window. It's in the Open and Save dialog boxes. It sits atop the window in Photos, Contacts, Notes, Maps, and Safari. (In Reminders and Calendar, you share using the ⊙ button instead, but the idea is the same.)

In each case, open the menu and choose **AirDrop**.

Now you see a list of nearby Macs, iPhones, and iPads that are set up to receive from you.

And what does "set up to receive from you" mean? See "How to Control Who AirDrops You" on the next page.

3. **Click the name of the person or machine you want to bless with your AirDrop package.**

   Actually, you can choose several people at once to save time. If the intended recipient's name isn't showing up, it's probably because she's got AirDrop set to **Off** or **Contacts Only** (and you're not in her Contacts app). Having read the box below, you now know how to advise her to change her setting.

   Another possible issue: AirDrop requires both Bluetooth and Wi-Fi to be turned on. Check the Control Center on the recipient's device.

   Oh, and the receiving machine has to be awake (turned on). You won't see its name if it's asleep.

   Finally, AirDrop isn't available on pre-2012 Apple machines.

At this point, the recipient sees a message containing a preview of whatever you're sending. It says, "[Other person's name] would like to share [the file's name]." The options are **Decline** or **Accept**.

HOW TO CONTROL WHO AIRDROPS YOU

Apple is fully aware that the world is full of idiots, and a lot of them are nasty.

When creating AirDrop, Apple's engineers had to ask: How do we prevent creating a hellscape of people bombarding one another with unsolicited AirDropped raunchy photos, obscene recordings, and terrorist manifestos?

The answer: You can control who may send you things by AirDrop.

To see this setting, click 🎛 on your menu bar to open the Control Center. Right there at top left, you can see your current AirDrop setting: **Off**, **Contacts Only**, or **Everyone**. To change the setting, click the current one.

(You can also adjust this setting in the AirDrop *window*, in the Finder. To see it, click **AirDrop** in any window's sidebar.

And on an iPhone or iPad, you change this setting in much the same way: Open the Control Center, long-press the **Wi-Fi** button cluster, and then tap **AirDrop**.)

Off means *nobody* can send you things by AirDrop. **Contacts Only** means that only people in your Contacts address-book app can send you things; presumably, you trust them at least a little. And **Everyone** means it's open season. Anyone within 30 feet of you can send you stuff by AirDrop.

Remember that you also must *accept* incoming AirDrop items; nobody can send you something without your awareness and permission. In other words, even the **Everyone** setting doesn't mean that your hard drive is going to fill up with crude files. ✦

If the receiver hits **Accept**, the transaction is complete. If you sent a file, it winds up on their Mac (in the Downloads folder) or in iOS's Files app. If you sent a note, map, web page, photo, or contact, it shows up in the corresponding app.

You can also AirDrop something to yourself—from your Mac to your phone, for example. In that situation, there's no **Accept** button to click; the incoming file or data morsel goes directly into the appropriate spot or app. In theory, you're not going to send upsetting photos to *yourself*.

> **TIP:** In the Finder, there's an alternative way to use AirDrop that's useful if you have a bunch of files to send. You start by opening the AirDrop *window*, which displays the icons of all the Apple machines within about 30 feet that are open to receiving stuff from you. To open the window, click **AirDrop** in the sidebar of any desktop window, or choose **Go→AirDrop**. To send a file or folder, just drag it onto the icon of the receiving device.

Choose who can see you.   *Click to see progress—and the ⊗ button.*

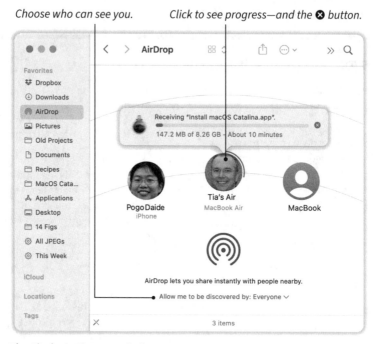

The Finder's AirDrop window

Public Folder Sharing

AirDrop is amazing, but it has its limitations. It works only on Apple devices. Also, it's a sender-driven process. Other people can't take files from your Mac on their own; you have to push the files to them while they're standing right next to you.

The Public folder method avoids those potholes.

Your home folder (page 110), and everyone else's, contains a special folder called Public. (To get to your home folder, choose **Go→Home** in the Finder.) Once you've turned on the master switch, any icons you drop into this folder are available to everyone else on the network, just like that. Nobody needs a password or an account; they're free to copy files out of your Public folder or even put files in.

To turn this feature on, choose **System Preferences→Sharing**, and then turn on **File Sharing**.

Master sharing switch   Add your own folders to share.

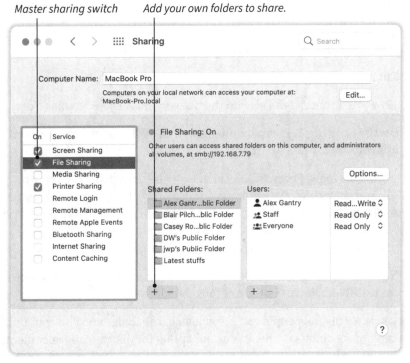

Turning on Public folder sharing

Now you can move or copy files and folders into your Public folder. Anyone else on the network can access them, using the instructions on page 340.

NOTE: Your Public folder starts out containing a subfolder called Drop Box. It lets other people on the network give you files. Read on.

"Any Folder" Sharing

The Public folder feature sure is simple. But it makes everything in your Public folder accessible to *everyone* on the network, including people who really have no business seeing it. It also requires that you *move or copy* your files from their original places on your Mac into the Public folder.

If you use the "any folder" sharing method, though, you don't have to move the files you want to share. You can share them right where they're sitting now. And using this method, you can grant various individuals different levels of access to your files.

These steps assume you've turned on **System Preferences→Sharing→File Sharing**, the Mac's master file-sharing on/off switch.

1. **Click the folder or disk you want to share. Choose File→Get Info.**

   The Get Info box appears, teeming with details about that disk or folder. If you can't see the **General** panel, click it to open it.

2. **Turn on "Shared folder."**

   Now you'll specify who else on the network has access to this folder.

3. **On the Sharing & Permissions panel, click the 🔒 and enter your administrator password.**

   If the Sharing & Permissions panel isn't already open, click the ❯ to expand it. Once you've entered your password, the controls are yours to change. See the little table at the bottom of the Get Info window? The Name column shows the names of individual account holders (or *groups* of account holders; see page 318).

   The Privilege column shows how much access each person or group has for this folder. It can say one of four things:

   Read only means people can see what's in the folder and copy out of it but can't make changes. This is a useful option for distributing documents to the whole company or family.

   Read & Write means people can take things out, put things in, or edit the documents inside.

   Write only (Drop Box) is an option only for folders and disks (not individual files). It means other network denizens can put files or folders *in*, but they can't open the folder or disk to see or copy what's inside. That's what you'd want if, for example, various writers are turning in their articles to you, the editor, or students are supposed to turn in their homework to you, the teacher.

Finally, **No access**—an option only for **Everyone**—means people on the network can see this file or folder's icon but can't open it, look inside, or even put anything into it.

4. **Add somebody's name and set that person's privileges.**

For the first victim, click the **+** button below the list. A panel appears, listing every account holder and group on this Mac (see Chapter 13). Click a name and then click **Select**. That person or group now appears in the Name column.

Now just tweak the pop-up menu in the Privilege column for this person: **Read only, Read & Write**, or **Write only (Drop Box)**.

> **TIP:** With one swift click, you can apply your painstakingly chosen set of names and privileges to everything *inside* this folder. That's a huge time-saver. To do that, click the ⬤ at the bottom of the Get Info box and choose **Apply to enclosed items**.

Give someone else access to this shared folder...

...and then specify how much access he has.

Get Info sharing panel

You've just shared the folder; congratulations. Close the Get Info window. At this point, your colleagues can begin to access the folder from their own Macs across the network.

Accessing Shared Files

You've now read about (or skimmed over) two ways to *set up* files to be shared communally on your Mac: the Public folder method of sharing files and the "any folder" method.

For your next trick, you're going to learn how to *access* shared folders. You're going to pretend you're the *other* guy—someone sitting across the office from you.

It's pretty simple, really. In any Finder window sidebar, click **Network**. The window now shows the icons of all computers on the network.

If you don't see the Mac you're looking for, maybe it's turned off, or maybe **File Sharing** is turned off in System Preferences on that Mac.

On really big networks—in really big companies—you may see *workgroups* listed here, too. Those are branches of your network tree. Just keep double-clicking until you find the machine you want.

In any case, now you should click the computer with the files you want. The Connect As dialog box appears, offering you three options:

- **Guest** is nice because you don't need a name and password to access the shared goodies. If you don't have a user account of your own on the Mac you're trying to access, this is the only option available to you.

  Unfortunately, if the owner of the other Mac hasn't turned on **Allow guest users to connect to shared folders** (in **System Preferences→Users & Groups→Guest User**), you'll get nothing but an error message.

*Click **Network**. Double-click a computer. Log in as though at that machine.*

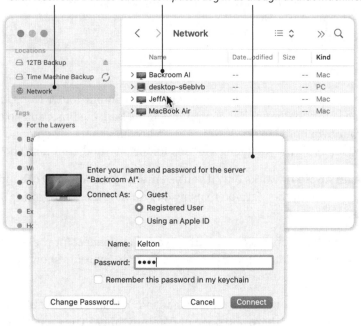

You're in! These folders are actually on the other Mac on the network.

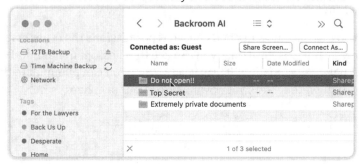

Shared Macs in the Finder

If guests *are* allowed in, when you click **Connect**, you see the Public folders of everybody with an account on the other Mac.

- **Registered User** is what you want if you have an account of your own on the other Mac. Enter your account name and password (for the Mac you're tapping into)—the same credentials you'd use to log in if you were sitting right there at that Mac.

Click **Connect**.

In the window that opens, you're looking at the Public folder icons for each account holder on that computer—Mom, Dad, Big Boss, and so on—and your own home folder, if you have an account there. If someone has shared folders with you using the "any folder" method, they also show up.

- **Using an Apple ID** works only if you've entered your Apple ID (iCloud address) on both the Mac you're using *and* on the Mac you're trying to access.

No matter which option you choose, when you click **Connect** (or press Return), an icon for the other Mac appears in the sidebar; its contents show up in the main window.

You can't delete or change anything in somebody's Public folder, but you're welcome to see what's there and open or copy anything. (The exception is the **Drop Box** folder. You're allowed to put stuff into this folder, but you can't see what else is inside.)

The fun you can have now depends on what permissions the other person has given you. If it was **Read Only**, you can see the files, open them, and copy them, but not delete or edit; if it was **Read & Write**, you can manipulate them exactly as though they were your own files on your own hard drive.

Disconnecting Yourself

When you've had enough fun working with the files on the shared folder, click **Disconnect** in the shared window to unhook from it.

Networking with Windows

Yes, you can share files between Macs and Windows PCs on the same office network. The question is, which way do you want the sharing to go? Do you want to share your Mac files with PCs? Or do you want to sit at your Mac and access files on a PC's hard drive?

Doesn't matter; you can do either one. It's a bunch of steps, and it's not always pretty, but it gets the job done.

Share Mac Files with Windows PCs

So you've decided to make some of your folders available to Windows PC people on the network. How generous of you! Here's how:

1. **Create a Mac account for the Windows person.**

   Yes, the Windows person needs an account on your Mac. See page 316 for instructions on creating one. The Windows person may never actually sit at your Mac to log in. But he will need a name and password to access the files you shared.

2. **Teach your Mac to speak Windows-ese.**

   Open **System Preferences→Sharing**. Click **File Sharing** and then **Options**. Turn on **Share files and folders using SMB.**

Your Mac's name on the network

*In Options, turn on **SMB** and then the Windows accounts that will have access.*

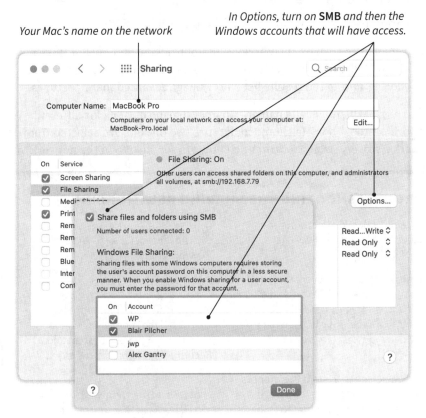

Windows File Sharing setup

Now you're looking at a list of the people for whom you've created accounts on your Mac; turn on the name of the account you created for the Windows person, and then click **Done**.

You're back on the Sharing pane of System Preferences. Notice (or change) the name of your Mac—its **Computer Name**.

At this point, the Windows person sits at his own machine, opens an Explorer window (his equivalent of the Finder), clicks **Network** in the sidebar, double-clicks the name of your Mac, and enters the name and password of his account on your Mac. It looks exactly like a Microsoft-ized version of the center dialog box in the figure called "Shared Macs in the Finder" (page 341).

A window opens up in Windows, revealing his own home folder on your Mac, along with other account holders' Public folders and any "any folder" folders they've shared.

Share Windows Files with a Mac

Now suppose the Windows person wants to share files with *you*.

His first job is to turn on the Windows file-sharing master switch, if it's not already on. The quickest way to find it: Click in the **Type here to search** box, type *sharing*, and in the results list click **Manage advanced sharing settings**.

He arrives at the musically named "Change sharing options for different network profiles" box. He should make sure two options are selected: **Turn on network discovery** (which lets other computers see the PC on the network) and

---

TROUBLESHOOTING THE SHARING RELATIONSHIP

In general, Mac-to-Windows sharing tends to require more fiddling than Mac-to-Mac sharing.

If, on the PC, you discover that your Mac's name isn't showing up, your first thought should be: "Are the Mac and PC on the same *workgroup*?" (In Windowsland, a workgroup is a branch of an office network.)

To confirm, open **System Preferences→Network** on the Mac. In the list at left, choose how you're connected to the network (**Wi-Fi,** for example), and then click **Advanced.** Click the **WINS** tab. In the **Workgroup** box, enter the Windows machine's workgroup name; it's probably either WORKGROUP or MSHOME.

(And how are you supposed to know which? Tell the Windows person to open **Control Panel→System and Security→System**; that's where his Windows workgroup name is revealed.)

Click **OK** and then **Apply**. Now the PC *should* be able to see the Mac. ✳

Turn on file and printer sharing (which lets him share files and folders). Click **Save changes.**

The rest of the process is exactly like sharing Mac files with Windows but held up to a mirror. For example, this time, the Windows person must create a user account for *you* on his PC.

Make sure the master sharing switches are on.

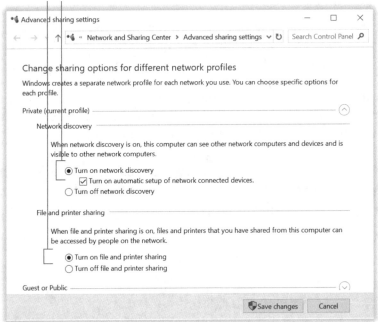

Share the file or folder on the Share tab.

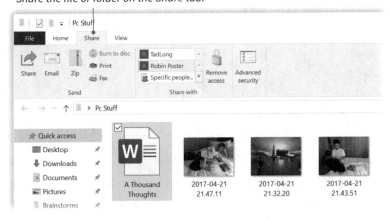

Sharing from Windows

Now, to share a certain folder with you, he opens its window and chooses **Share→Specific people** and chooses your account name. (If you're not already in the list, he types your account name and hits **Add**.) He adjusts your permission—**Read** (you can see it but can't change or delete it) or **Read/Write** (you can edit or delete)—and then hits **Share**.

On your Mac, you can now see the PC listed under Network in any Finder window's sidebar, listed alongside whatever Macs are also shared.

> **NOTE:** On a big network, you may have to click **All** to see the icons of its *workgroups*—network clusters—before you see the names of the actual PCs. Double-click the workgroup name you want.

Open it, enter the name and password for the Windows account the Windows person set up for you, and enjoy full access to the shared folders and files. They behave exactly as though they're sitting right there on your Mac.

> **TIP:** If you value your time, turn on **Remember this password in my keychain**. That way you don't have to enter your Windows account credentials the next time you want to access the shared stuff.

CHAPTER FIFTEEN

Fourteen Mac Annoyances

T he Mac may be the world's least annoying computer, but let's face it: It's still a computer. Over the years, Apple has employed thousands of programmers, each leaving a little stamp on macOS. Eventually, you'll come across things that don't work right, that frustrate you, or whose design baffles you.

May the following guidance spare you a weekend or two of Googling and swearing.

Lockups and Freezes

Computers, like you, occasionally become overwhelmed, overstressed, and confused. In the Mac's case, that can mean unresponsiveness. Maybe it's just one app that goes catatonic, or maybe it's your whole Mac.

If this happens, you're not required to throw the Mac away and buy a new one.

An App Locks Up: Force-Quit

If an app goes frozen on you, or your cursor changes to a colorful pinwheel (the Spinning Beach Ball of Death, as insiders know it), what you want to do is *force-quit* it, or exit it with brute force.

Choose →**Force Quit**. Or press the classic keystroke, Option-⌘-Esc. In either case, the Force Quit dialog box appears. Click the name of the broken app, and then click **Force Quit**. That's it: The app is no longer locked up, or even open. (Page 139 shows force-quitting in action.)

The Whole Mac Gets Lockjaw

Every now and then, things get *really* bad. The whole machine seems locked up. Nothing responds to your mouse clicks or keystrokes.

In that situation, you can't open the menu to reach the **Force Quit** command, and no keyboard shortcut registers with the Mac. You have no choice but to restart the entire computer.

To do that, hold down the power key for six seconds. The screen goes black, you lose your work in any documents you hadn't saved, and the Mac starts up again.

Screen Brightness Keeps Wobbling

Apple actually thought this was a feature you'd *like*: The Mac has an ambient-light sensor. When the room is very bright, the Mac makes the screen bright, too. When the room is dark, the Mac dims the screen. It's all supposed to improve visibility in various light conditions without your having to fiddle with anything.

In practice, though, the auto-dimming stuff can be distracting and a little overzealous. You may find the screen brightening and dimming seemingly randomly as you're trying to get work done.

To turn off the automatic dimming, open **System Preferences→Displays→ Display** and turn off **Automatically adjust brightness**. You can always adjust the brightness yourself, as needed, in the Control Center ().

Nutty Things Happen When You Click

Apple is very proud of the Force Touch trackpads on its recent laptops. These trackpads don't just know when you've clicked; they know *how hard* you've clicked.

If you click normally, and then press *harder,* you feel a distinct second click, as though you've pushed the trackpad down to a lower level.

Force-clicking can summon certain informational bonuses, as described on page 71. But force-clicking can also drive you crazy. You may wind up producing pop-up panels of information accidentally, not realizing that you've clicked to the secondary level of pressure.

If you're not using force-clicking, you may as well turn it off so you don't fire it up by accident. You do that by unchecking **System Preferences→Trackpad→ Point & Click→Force Click and haptic feedback**.

Phantom Clicks on the Trackpad

This can be a tricky one to diagnose. Since a click is invisible, all you may detect are the *effects* of phantom clicks: windows that abruptly move or scroll by themselves, buttons that seem to click themselves, weird dialog boxes that appear unbidden, error beeps you can't figure out.

Somehow, you've turned on **Tap to click** in **System Preferences→Trackpad→Point & Click**.

May become your best friend—or drive you crazy

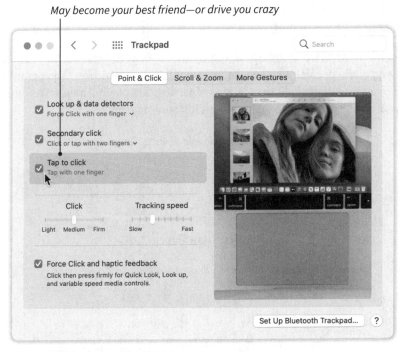

Tap to click

It's actually a cool feature, if you're ready for it. It means you can click things just by *touching* the trackpad instead of actually *clicking down* on it. **Tap to click** means you can click silently, which is useful if, for example, someone is trying to sleep next to you. Or maybe you just like the gentle, fluid, nearly effortless action of touching the trackpad instead of clicking.

But if you don't know what's happening, it's chaos. Every time one of your fingers brushes the trackpad, the Mac thinks you're trying to click—at whatever random spot the arrow cursor happens to be—and bizarre things can result.

Sound Keeps Playing from Somewhere

It's enough to drive you around the bend: Some stupid app or web page is playing video. You're hearing the audio track, and you dearly wish it would stop, but you can't seem to find the source!

Fifteen of the built-in Mac apps can play sound. Some are *designed* to play sound, like Books (for playing audiobooks), GarageBand, Music, Podcasts, and Voice Memos. Some are designed for making videos (iMovie, QuickTime Player), and some can play videos as part of their other duties (Keynote, Mail, Messages, Pages, Photos). If any of those is the source of the sound, it's probably not hard to find; you'd remember having clicked ▶ on some video.

Most often, what's playing is a web page in one of your browser tabs. Those are especially pernicious because (a) you might have a dozen tabs open, and (b) some websites take it upon themselves to start playing a video unbidden. How are you supposed to guess where it's coming from?

Click to shut up audio on all tabs—or hold down for a menu of audio tabs.

The icon also appears on the offending tab itself. Click or hold down.

Mute tabs

Apple has provided two weapons against that obnoxiousness. First, choose Safari→Preferences→Websites→Autoplay, and notice the **When visiting other websites** pop-up menu. You definitely want it to say **Never Auto-Play** (or, at the very least, **Stop Media with Sound**, which lets videos auto-play only if they're silent).

> **TIP:** You can override this setting on a website-by-website basis, as described on page 211.

But what if it's too late for that? What if you've got a bunch of windows or tabs open *right now*, and something's playing, and you can't find it?

Safari's got you. Look for the ◀)) in the address bar. That's the **Mute All Tabs** button, which is Apple's gift to your sanity. There's also a black ◀)) on the tab that's actually doing the playing. It's not just an indicator; it's also a Shut Up button.

An Open App Goes Missing

This one'll drive you batty. You've opened an app. You know it's open. Maybe you can even hear it, because it's a video, a song, or a call you're on. But you can't *find* the darned thing. You choose the app's icon on the app switcher (page 136)—and the window you want isn't there. You even enter Mission Control (page 134), whose entire purpose is to show you everything that's happening on your Mac. And the window isn't there, either!

As techies like to say, their voices dripping with sarcasm: It's a feature, not a bug.

What's happened is that you've put one of your apps into *full-screen mode*. You can read more about this mode on page 90, but the gist is that you clicked the green ❷ button at the top left of a window, and it expanded to fill your screen, edge to edge. Its menu bar, scroll bars, and other junk are hidden, all in the name of maximizing your working area.

So far, so good.

But what full-screen mode *also* does is create a second, entirely *virtual* screen, off to the right of your regular screen! You now have two of what Apple calls Spaces. If you're technically proficient, you might even like creating Spaces, because it's like having an external monitor (or two, or three, or more). In Space 1, you might have your social media apps. In Space 2, your email and Messages. In Space 3, Safari in full-screen mode. In Space 4, the work you're actually supposed to be doing today.

Space 1:
Your work

Space 2:
Email

Space 3:
Facebook and Twitter

A map of your Spaces

It's also possible to have different documents from the *same app* running in different Spaces.

If it sounds as though it could get confusing, then you've got a good grasp on the concept.

So how do you move from Space to Space? If you knew that, the Case of the Missing App Window would be solved!

Apple gives you many different ways:

- **On a trackpad,** swipe horizontally with three fingers.

- **On a Magic Mouse,** swipe horizontally with two fingers.

- **Enter Mission Control (page 134).** At the very top of the screen, you see the names of your different Spaces. Click the one you want.

- **Use the arrow keys.** Press Control-◄ or Control-► to slide to the previous or next Space.

When you use any of these techniques, your current Space slides away, and the new one slides in.

Your reaction to this explanation is probably somewhere on the continuum from "Oh, cool! That will help me get organized," to "That's ridiculous! I think I'll just avoid using full-screen mode from now on." Adjust accordingly.

You Can't Open a New App

So. You're trying to open a new app, and macOS flat-out refuses. It primly tells you it comes from an unknown source, and you're not allowed to open it. Here's the explanation—and the workaround.

Malware—software designed to hurt you instead of help you—often takes the form of apps that take over your Mac or invade your privacy. Maybe you downloaded one of these apps yourself because you thought it was innocent, or maybe you weren't even aware you downloaded it.

Either way, the Mac security feature called Gatekeeper makes that bad-guy tactic difficult or impossible. Its job is to prevent you from opening apps from sources that Apple hasn't determined to be safe.

> **NOTE:** All apps from the Mac App Store (page 131) are safe. Apple has individually vetted each one to confirm that it has no ulterior motives. Gatekeeper watches over apps that come from *other* sources.

To check it out, open **System Preferences→Security & Privacy→General**. Click 🔒 and enter your password to unlock this panel. At the bottom of this screen, you see two **Allow apps downloaded from** options. That's Gatekeeper.

The word "Gatekeeper" doesn't appear here, but this is what's blocking your access to apps or software companies that Apple hasn't approved.

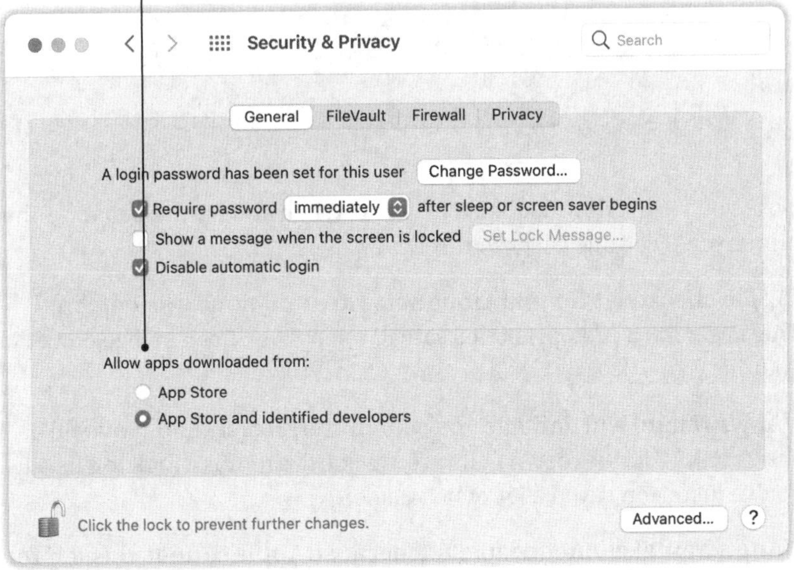

Gatekeeper settings

These are your options:

- **App Store.** Apple has tested and vetted every single app on the Mac App Store. If you stick to these apps, it's impossible for you to get malware.

- **App Store and identified developers.** This setting lets you install both App Store programs *and* apps from software developers that have registered with Apple. (Translation: No Eastern European teenagers with too much time on their hands.)

Now, Apple doesn't hand-inspect every app from these companies. But if anybody reports that one of these apps turns out to contain malware, Apple adds it to the Gatekeeper blacklist, which it updates daily. At that moment, nobody else using Gatekeeper will be able to install it.

In Big Sur, Gatekeeper has grown even more strict. It makes life very clumsy if you try to open any app that hasn't been *notarized* (checked out by Apple in advance).

Notarization notice

Suppose you've just downloaded a new app. When you try to open it for the first time, you'll see one of these three messages:

- **"Apple checked it for malicious software and none was detected."** This is a notarized app, and it's safe to open. Click **Open** and go on with your life.

- **"[App's Name] will damage your computer. You should move it to the Trash."** This is known malware. Click **Move to Trash**, and don't argue. You've just been saved a lot of headache.

- **"[App's Name] cannot be opened because Apple cannot check it for malicious software."** This app hasn't been notarized. You can open it, but only if you're really confident it's safe—and really patient.

To go ahead, right-click or two-finger click the app's icon; from the short-cut menu, choose **Open**. Click **Cancel**. Then right-click or two-finger click the app's icon *again*, and choose **Open** *again*. This time—yes, only the second time—an **Open** button appears. Click it to open the app, overriding the Mac's advice.

Goofy? Yes. But the app will run normally from now on.

Some App Auto-Launches Every Day

There should be a Mac Owner's Bill of Rights. And somewhere in the top 10, it should say: "No program opens unless I ask it to."

That would rule out apps that, once you've installed them, decide for themselves to auto-open every time you use your Mac.

If those self-important apps are getting on your nerves, you can squash them. Open **System Preferences→Users & Groups→[your account]→Login Items.** Here's the list of all apps that you've asked to auto-open—and those that have chosen that honor for themselves. Click the offending app and then click the − button to remove it from the startup list.

Stop self-opening, you fool!

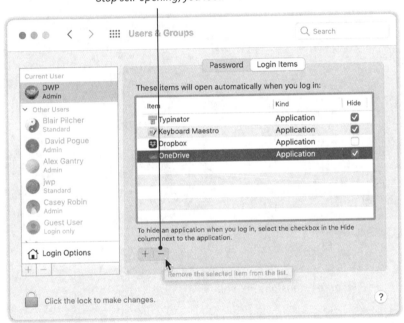

Stopping a self-opening app

The Wrong App Opens

When you double-click a PDF document on the Mac, Preview opens to show it. Not Safari, not Pages, not Notes, even though all those apps are also capable of displaying PDF files.

Which is all well and good if Preview is the app you *wanted* to open. But sooner or later, you'll find yourself in some situation where you double-click a file, and it opens into an app that you didn't want. Maybe you've got Adobe Photoshop, which is fantastic for editing photos—but when you double-click a photo in the Finder, it opens in Preview instead. Now what?

Behind the scenes, almost every document in Macdom has a three- or four-letter *filename extension*—a code that tells the Mac what kind of document it is. Is it a photo (.jpg)? A PDF document (.pdf)? A text file (.txt)? A Microsoft Word file (.docx)?

> **NOTE:** Microsoft Windows identifies documents using these filename extensions, too. You can email a .jpg, .pdf, .txt, or .docx document to somebody who's not a Mac person—and they'll have no problem opening it.

Which app do you want to open this JPEG file? So many options!

Filename extensions in Get Info

Usually you don't see these suffixes because the Mac hides them. Apple considers them geeky-looking and intimidating.

> **NOTE:** If you'd *like* to see them, choose **Finder→Preferences→Advanced**, and turn on **Show all filename extensions**. Now your documents display their previously hidden suffixes.
>
> You can hide or show these suffixes one icon at a time, too. Highlight an icon (or several) and then choose **File→Get Info**. In the resulting Get Info window, open the **Name & Extension** panel and turn **Hide extension** on or off.

Now, some documents are openable by only one app. You might use MustacheDesigner Pro 2.0 to create 3D printable mockups of your facial adornments, resulting in documents called things like *New Handlebar.mdp*. When you double-click that file, the MustacheDesigner Pro app opens, because no other app would know what to do with an .mdp file.

Many other document types, though, can be opened by a variety of apps. Just about any app worth its code can open a JPEG photo file, for example—Preview, Safari, QuickTime Player, Notes, Reminders, Pages, Numbers, and many more. How does macOS know which app should open when you double-click the file?

That's entirely up to you.

Reassigning One Document, One Time

To set up a document to open in one particular app, only one time, select its icon and then choose **File→Open With**.

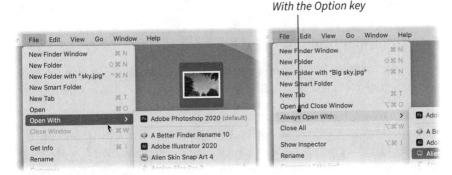

With the Option key

The Open With command

You get a menu of apps capable of opening your document. It offers choices like these:

- **(default).** This is the app, listed at the top, that *usually* opens this kind of document.

- **[other apps].** These other apps can open your document, too.

- **App Store.** The Mac takes you straight to the Mac App Store, where, in principle, you should be able to find an app that can open your oddball file.

- **Other.** What if you have an app that *you* know can open this document, but macOS *doesn't* know? Use the **Other** command to locate and choose the app you want.

Choose the app you'd like to open this one document right now.

Reassigning One Document for Good

If you use **Open With** as just described, you haven't changed your document's *future* behavior. The next time you double-click it, it still opens with the **(default)** app.

But what if you want to reassign it to a new app for good?

The steps are almost the same (**File→Open With**)—but once the shortcut menu appears, press the Option key. Before your eyes, the **Open With** command changes to say **Always Open With.**

> **TIP:** You can use this trick on a big group of selected icons, thereby reassigning them all simultaneously.

Reassigning a Document Type

You've just read the instructions for teaching the Mac that you want to open *one particular document* into a different parent app, every time you open it.

But there may come a time when you want to reassign *all* documents of that type. Maybe you want all .mp3 music files to open in QuickTime Player instead of the Music app, because QuickTime Player is quicker to open and easier to use. Or maybe you want all PNG screenshots to open in Photoshop instead of Preview, now and forever.

To associate a particular filename extension to a new app, start by clicking *one* of them—for example, one JPEG photo. Choose **File→Get Info.** In the Get Info box, expand the **Open with** section, if necessary. Choose the name of the app you want to open this kind of document from now on.

But here's the key step: Click **Change All** beneath the pop-up menu. Confirm by clicking **Continue** or pressing Return.

The surgery is complete. From now on, double-clicking any document of that type opens it in the newly selected program.

Doesn't it feel good to take control of your world?

Notification Bombs

Notifications are the rounded message boxes that appear at the top right of your screen. They appear to let you know—well, *everything*, and that's the problem.

You get a notification when an email comes in, when a text message arrives, when there's been a charge on your Apple Card, when there's a new Facebook or Twitter post, when a calendar appointment arrives, when someone changes your calendar, when anything goes wrong with Time Machine, when there's been a change to your Dropbox contents, when someone wants to play a game with you, when Maps needs your attention, when someone edits one of your shared Notes or Reminders ... and on and on.

Pretty soon, it's dinnertime and you've gotten no work done today. You spent the entire day dismissing notification bubbles.

You can shut down those notification bombardments either by turning on Do Not Disturb (page 48) or by shushing the notifications from the noisiest apps (page 45).

Running Out of Disk Space

The hard drive in most Mac models is more or less permanently installed. You're expected, upon buying the Mac, to somehow *anticipate* how many files, photos, videos, music, apps, and documents you'll ever accumulate.

Forget it. Sooner or later, every hard drive fills up, and that's a *huge* problem. It's not just that you're running out of room for new files. It's that weird stuff starts happening, like random error messages, apps quitting suddenly, mysterious lags switching from one app to another, features just not working.

That's because the Mac needs a big swath of free hard drive space (at least 10% of the total size) for its own operations. Behind the scenes, the Mac is constantly picking up and putting down pieces of files and apps as it juggles its tasks. If space gets tight, all that starts to take a painful amount of time— and if it gets *really* tight, your apps start to suffocate, and really bad things happen.

The Mac's Storage-Reporting Center

How full is your drive at this moment? Fortunately, macOS offers a handy gauge—and a bunch of one-click ways to free up some space.

Choose →**About This Mac**→**Storage**. Here you get a handsome, color-coded bar chart that shows how full your drive is at this moment (see "Managing your storage" on the facing page). You can point to a section of the chart without clicking to see what it represents: Apps, Documents, Messages (all the pictures, videos, and other stuff people have sent you), and so on.

Even sweeter: The **Manage** button opens a world of utilities for reclaiming drive space—by deleting, for example, old files or orphaned support files for apps you no longer use. Here are the ways it proposes to tackle your capacity problem:

- **Store in iCloud.** Apple, in its generosity, is prepared to offload some of the most space-greedy files from your hard drive—onto its own servers online.

NOTE: As you know from page 105, iCloud storage costs money. You get 5 gigabytes of free storage—about a thimbleful of data—and then you have to pay Apple monthly for a bigger allotment of space. It's $12 a year for 50 gigabytes, $36 a year for 200 GB, or $120 a year for 2 TB. Read about what that gets you, ponder the long-term cost, and then make your decision.

The first option here, **Desktop & Documents**, moves the contents of these two folders from your Mac onto your iCloud Drive instead. Most people store most of their files, after all, either on the desktop or in the Documents folder. Therefore, this option not only frees up a huge mass of disk space, but it also makes the contents of those folders available to you from any Apple gadget: other Macs, your iPhone or iPad, or even the web, at iCloud.com. See the box on page 362 for details.

The other option here, **Photos**, addresses the other biggest space glutton in most people's lives: their Photos stash. All the pictures and videos they've ever taken can add up to many, many gigabytes' worth of files—space that you get back when you turn this option on.

This feature is described on page 286; for now, the takeaway is that this space-freeing process kicks in silently and automatically, and only when necessary—when your drive starts to get dangerously full—and in no way inconveniences you. The photos *seem* to still be present on the Mac, because Photos retains copies that are just big enough to fill your screen, for your perusing pleasure. (The originals are many times bigger.)

In any case, turn on the checkboxes of the options you'd like to enable, and then click **Store in iCloud**. Marvel as, a few hours later, your hard drive shows a lot more free space.

Click to view the Mac's built-in file-purging tools.

*Inspect each category of huge files, which come with **Delete** buttons.*

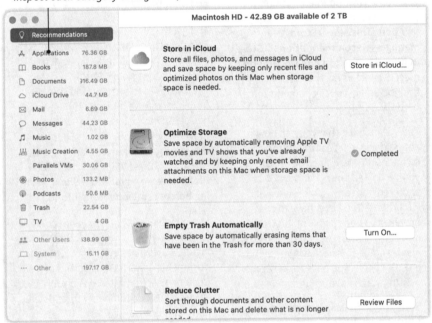

Managing your storage

- **Optimize Storage.** If you click **Optimize,** you're offered just one option: to have the Mac auto-delete movies and TV shows (that you've bought from Apple's TV store) after you've watched them.

> **NOTE:** Weirdly, you can't turn this feature *off* again in the same place. If you want to turn the feature off, open the TV app, choose **TV→Preferences→Files,** and turn off **Automatically delete watched movies and TV shows**.

This feature, too, kicks in only when your drive gets low on space (under 10 GB). And, of course, you *own* those shows and movies; you can always re-download them at no charge.

HOW "DESKTOP & DOCUMENTS IN ICLOUD" WORKS

Apple's mighty ecosystem of software and services offers a bunch of ways to transfer stuff from your overcrowded hard drive to Apple's online storage servers. You save space on your Mac, yet the offloaded files *look and feel* as though they're still in place on your drive.

The Desktop & Documents option, though, is complex. If you turn it on (in **About This Mac→Storage→ Manage→Recommendations→Store in iCloud→Desktop and Documents**), macOS transfers everything on your desktop and in your Documents folder onto iCloud.com.

The advantage is that you can now access these most common storage locations on every Mac you own, every iPhone or iPad, every Windows PC, and even from any web browser, at iCloud.com. You never have to copy or send files back and forth between machines; almost everything you ever need is grabbable from wherever you are. (Of course, you have to pay Apple monthly for the iCloud storage.)

But some complicated and probably worrisome things happen when you use this feature, and you should be warned.

First, all the icons that used to be on your desktop are *gone*. They've been swept into a new folder called "Desktop - iMac" (or whatever your Mac's name is). Everything in your Documents folder disappears, too. *Those* icons are now in a folder in your Documents folder called "Documents - iMac."

Apple wants to distinguish what's on the desktop of *this* Mac from what's on the desktops of your *other* Macs. If you've turned this feature on for two Macs, in other words, you might see folders on your desktop called *Desktop - iMac* and *Desktop - MacBook Pro.*

The second big change: In the sidebar of every Finder window, you get new entries (in the iCloud category) for **Documents** and **Desktop**. Click one to see all the stuff that *used* to be in your Documents folder and on your desktop.

And get this: If you drag an icon out of Documents or off your desktop, you get a warning: "Are you sure you want to remove this file from iCloud? This item will be moved to this Mac and deleted from iCloud Drive and your other iCloud devices." That makes sense—you're moving back to your Mac and *out* of the iCloud universe—but it's alarming. (You can turn off these warnings in **Finder→ Preferences→Advanced→Show warning before removing from iCloud**.)

Something equally unnerving happens if you ever turn *off* iCloud Desktop & Documents. A message says: "If you continue, documents on your Desktop and in your Documents folder will be visible in iCloud Drive only. No documents will be deleted from iCloud Drive."

And now you look inside your Desktop and Documents folders—the ones listed under iCloud in the sidebar—and they're *empty*! What the what?

Remain calm. All those files are still on your Mac—but now they're in the iCloud Drive folder, in folders called Desktop and Documents. You're free to open those folders and drag their contents back.

The point is that this can be a useful and powerful feature. But because it involves all your important files seeming to disappear completely—more than once!—you should know what you're in for. ✦

- **Empty Trash Automatically.** Some people put stuff in the Trash (page 119) and then never *empty* the Trash. Over the years, their hard drive may get dangerously full—because there are 600 gigabytes of stuff rotting in the Trash. If you click **Turn On** (and then **Turn On** again), the Mac will auto-delete files you've put in the Trash after they've been there for a month.

- **Reduce Clutter.** If you click **Review Files**, you get to see a list of all your documents, with handy tabs like **Large Files**, **Downloads** (usually lots of installers and other stuff you've downloaded and then forgotten about), **Unsupported Apps** (that is, old, 32-bit apps that don't run in Big Sur), and **Containers**. The idea is to give you easy access to the biggest and most useless files that are eating up your drive space—and let you delete them with a quick click of the **Delete** button (and then a confirmation click).

> **TIP:** To see what something is, click it once and then press the space bar for a Quick Look at it (page 103).

The last tab, **File Browser**, is just a column-view list of everything on your drive, biggest folders listed first. Again, the idea is to make it easy for you to spot space-eaters and then delete them using the **Move to Trash** button.

In the Manage Storage window's sidebar, you get a list of other large-file categories: **Applications, Books, Podcasts, TV,** and so on. Click to see all the files you've got of that file type, along with how much space they're occupying. The **Delete** button is right there.

Music Creation refers to the digital instrument sounds and music-instruction videos in GarageBand, Apple's music-creation program; they take up huge amounts of disk space. (You can always re-download them later.)

Mail and **Messages** show you the grand total of all your email and **Messages** attachments. **Music** offers a list of your music videos, ready for deleting. You can also review what's in your **Trash**, see how long things have been in there, and delete them.

A couple of items appear in gray, meaning you can see how much space they're occupying but can't do anything about it. They include **Other Users** (how much space other account holders' home folders occupy) and **System** (macOS itself).

Optimize Mac Storage

After that exhausting tour of macOS's space-saving features, you may be ready for a nap when you discover that they're not the end of it.

Remember that business of iCloud Photos, which auto-stores the original, jumbo photo files online to free up space on your Mac as your drive nears fullness? (If not, see page 286.)

Well, the Mac can perform exactly the same trick with regular files in the Finder. That is, when you start running low on space, the Mac offloads files to iCloud, starting with the oldest ones. Each offloaded file's icon remains on your Mac, marked with a ☁ badge. When you double-click that icon, the Mac downloads and opens it seamlessly and instantly, as though that file had never left.

For some reason, the switch to turn this feature on isn't in the same place as the other iCloud storage options described here. Instead, to turn it on, open **System Preferences→Apple ID**, and turn on **Optimize Mac Storage**. Here again, using this feature may require paying Apple for some expanded iCloud space. But, here again, the whole thing is dreamily automatic and invisible. You're spared the headache of "storage space is running low" messages forever (or until you stop paying).

Messages in iCloud

You might not think that text messages would eat up much disk space—and you're right.

But remember that what makes Apple's version, iMessages, so special is that they're *not* just bits of text. They can be audio recordings, videos, photos, files, web links, Memoji—all kinds of multimedia goodness that, while superb for self-expression, eat up disk space like crazy. Messages never deletes anything you send or receive (unless you've changed its factory settings), so over the years, this stuff really piles up.

You're not stuck in that situation, however. You can, if you wish, offload all that Messages history to iCloud storage space rented from Apple.

Doing so has three immediate advantages. First, you instantly save many gigabytes of space—not just on your Mac, but also on your iPhone, iPad, and other Macs! That's right: Until you turn this feature on, every one of your Apple gadgets maintains a *duplicate* copy of every photo, video, and audio clip you've ever sent or received in Messages!

Second, there's no more waiting for your phone to "catch up" with all the messages you exchanged using your Mac during, for example, a plane ride.

Third, when you get a new Apple machine, Messages presents to you your entire history of text messages and attachments. (Without Messages in iCloud, you'd see only your current and future messages. You couldn't scroll back into the past.)

If that sounds good—and if you don't mind the amount of iCloud storage space this feature will consume—open **Messages**→**Preferences**→**iMessage**, and turn on **Enable Messages in iCloud**. Your Mac begins backing up your Messages history, which can take quite a while.

You have to turn this feature on separately on each device; on the iPhone, you do that in **Settings**→**[your name]**→**iCloud**→**Messages**. The phone or tablet has to be plugged into power and on a Wi-Fi network to begin the offloading to iCloud.

Add to Your Built-In Storage

It's difficult, but not impossible, to replace your Mac's current drive with a bigger one. On all models except the Mac Pro, it's a job for a trained technician. And, of course, it costs money—although unlike the iCloud solutions, it's just a one-time payment.

> **NOTE:** On laptops, the Mac mini, and some iMacs, it's not even a "drive"—it's a tiny memory board covered with chips. It is, in other words, a solid-state drive, or SSD, which is costly but gives you much better battery life and ruggedness than a spinning disk would.

You can also buy an external hard drive. It plugs into your USB or Thunderbolt jack, whereupon you can offload files and folders from your overloaded Mac drive to your nice, big, empty external one. Surely there are files you don't actually need to have present on your day-to-day Mac.

Apps Reopen Unbidden

Here's another feature Apple hoped would save you time—but may actually bring you bafflement.

When you log out or turn off the Mac each day, macOS tries to remember which apps you were using, which windows were open, and how they were positioned. Then, the next time you log in, they all reopen automatically so you can dive right back in to whatever you were doing.

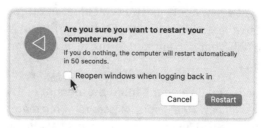

Restart options

If you'd prefer to arrange that yourself each time you log in, no problem: At the moment you choose the **Restart** or **Shut Down** command from the menu, just turn off **Reopen windows when logging back in**. And that's the end of *that*.

> **NOTE:** A similar feature tries to reopen all the windows you had open in each app the next time you open that app (whether or not you restarted the Mac). You can make it stop in **System Preferences→General**; turn on **Close windows when quitting an app**.

Losing Your Downloads

Finding files you've downloaded from the internet once they're on your Mac—apps, photos, PDF files, what have you—has been a headache since the dawn of personal computing. After all: When you save a new document from an app, it *asks* you what folder you want to save it in. But when you download something in your web browser, nobody asks you where to save it—or tells you where it went.

Apple tried to fix this problem years ago by creating the Downloads folder, which occupies a place of pride in your home folder. It's also listed in the sidebar of every Finder window. (If you don't see it there, open **Finder→Preferences→Sidebar** and turn it on.)

On top of all that, the ↓ button at the top of the Safari window (which appears after you have, in fact, downloaded something) opens a menu of recent downloads; see page 207. Each one has a ⊙ button that takes you directly to the downloaded file, wherever it wound up. (Hint: It's probably the Downloads folder.)

The Downloads folder also, by the way, contains files that people have sent to you by AirDrop (page 333).

In short, macOS is full of gigantic neon arrows that direct you to the Downloads folder. That doesn't mean, however, that you can't choose a different place. Plenty of people prefer downloads to wind up right on the desktop, where they're impossible to miss.

Fortunately, in Safari, you can choose **Safari→Preferences→General** and use the **File download location** pop-up menu to choose a different landing place for your downloaded goodies.

And if even *that* doesn't do it for you, there's always Spotlight (page 120).

The Mac Won't Start Up

If you live a clean life and make the proper sacrifices to the technology gods, you'll never experience *this* one. It's not so much an annoyance, actually, as it is a *freaking nightmare that ruins your weekend.*

When something goes seriously wrong with your Mac, it may not even start up. You may get a blinking question-mark folder icon on the screen, or a *kernel panic*—a crazy techno-babble paragraph that fills your screen in every language it can think of.

Often, restarting the Mac solves the problem. Or maybe you've just made some hardware change—installed a new drive, installed new memory, or attached some new equipment—and uninstalling it fixes the problem.

If those aren't the solutions, it's time to dive into some of the Mac's potent Deep Troubleshooting tools.

Safe Mode (Safe Boot)

When a computer starts up, it loads a long list of invisible background apps— the stuff that establishes your internet connection, monitors security, and so on. (That's why starting up takes so long.)

But if some piece of software, some font, or some startup process goes wrong, the Mac won't ever finish starting up.

To find out if that's the problem, press the Shift key as the Mac is starting up. Keep it down from the startup chime until you see the words "Safe Boot" on the login screen. You've just entered Safe Mode.

Entering Safe Mode runs a hard drive check, turns off all kinds of startup-software pieces, turns off any fonts you've added, and prevents your login items (page 322) from opening. But at least you've gotten your Mac *running.* Now, through process of elimination, you can figure out whatever font, app, or software driver was causing the problem.

Recovery Mode

But what if the problem is the hard drive itself? Or what if macOS itself has gotten corrupted?

Fortunately, you've got a backup. When you installed Big Sur, macOS quietly turned a chunk of your internal drive into a simulated drive called Recovery HD, complete with its own disk icon. It has its own copy of macOS, so you can start up your Mac even when your *main* macOS copy has gotten munged.

The Recovery HD also contains some first-aid apps that can fix your drive or software glitches, restore your backup files, browse the web, and even reinstall a clean copy of macOS.

To start up the Mac from the Recovery HD drive, do one of these:

- **As the Mac starts up,** press ⌘-R. When the Apple logo appears, release the keys.

- **As the Mac starts up,** hold down the Option key. When you see the Startup Manager, let go. Click the up-arrow button below the Recovery HD icon.

> **TIP:** Holding down the Option key during startup is the universal technique for choosing a different startup drive. Maybe, for example, you've attached an external drive with its own copy of macOS on it. This is how you tell the Mac to start up from *that* drive instead of the built-in one.

Startup Manager and Utilities

Log in with an administrator account (page 317). You arrive at the Utilities screen. On this screen, four emergency tools await you:

- **Restore From Time Machine Backup.** If you've got a Time Machine backup (page 64), it's payoff time. This option completely erases your hard drive—and then restores everything that used to be on it, including programs, data, settings, and macOS itself, from the backup. It takes a couple of hours, but it's an incredible safety net.

- **Install macOS.** This option downloads a new copy of macOS Big Sur from the internet; it's over 12 gigabytes, so it won't exactly be instantaneous.

- **Get Help Online** opens Safari to a page that offers Recovery Mode instructions.

AN EASY WAY TO FIX THE DISK

Since this chapter has veered into full-on troubleshooting, here's something you should know: A huge number of mysterious glitches, freezes, and hiccups owe their existence to tiny flaws—hardware or software—on your drive.

Fortunately, Apple has equipped you with Disk Utility, a highly skilled but easy-to-use disk-repair app. It's in your **Applications→Utilities** folder. (You can also open it from the Recovery mode described on these pages.)

At the left side of the screen, Disk Utility lists all the drives it can find. Click the disk or disk partition you want to fix, click **First Aid**, and then click **Run**. (A *partition* is a chunk of a drive that software treats as a separate drive, with its own name and icon on your Mac.)

Often, Disk Utility finds things to fix, fixes them, and then congratulates you with "The volume 'Macintosh HD' appears to be OK."

If it encounters something more serious, you'll have to take the Mac in for repair—or, if you're up for it, try a disk-repair program like DiskWarrior (alsoft.com). ✦

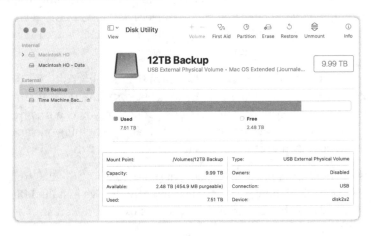

- **Disk Utility** is the Mac's hard drive diagnostics and repair program. What you'll want to do with it, in most cases, is to repair your hard drive. See the box "An Easy Way to Fix the Disk."

Internet Recovery

That secret Recovery HD drive is a masterstroke, and it's saved untold thousands of people from having to take their sick Macs in to repair shops. But what if the Recovery HD drive itself is somehow screwed up?

Or what if you've installed a new, empty hard drive—you've upgraded to a bigger one, for example—that doesn't have Recovery HD on it?

In those situations, you can use *Internet Recovery*. That's basically an online version of the Recovery HD, offering the same Utilities screen. Yes, you can actually start up your Mac from a drive that's thousands of miles away and use it to fix your *actual* drive or install macOS onto it.

To use Internet Recovery, you'll need an internet connection with decent speed. Turn on the Mac; as it's starting up, hold down ⌘-R. When the screen asks you to choose a Wi-Fi network, let go. Sign into a Wi-Fi network.

After a few moments, the Mac downloads a simulated disk. It's got a set of system software that brings you to the Utilities screen described in the previous section.

More Troubleshooting Help

The world's best source of troubleshooting information is Google.com. A single Google search includes Apple's own official help articles, plus thousands of articles and how-tos from other sources.

The trick is to specify your exact situation when you're searching. You might try, for example, *can't find date on menu bar clock mac big sur*. Or *microsoft word for mac add page numbering*. It's amazing how quickly you can find exactly what you're looking for.

If you'd like to limit your quest to Apple's own tech-support site, visit apple.com/support. It's full of manuals, articles, software updates, and frequently asked questions. At discussions.apple.com, you can find hundreds of thousands of conversations among fellow Mac fans helping one another out. You can post your own questions here, too.

And, yes, there is an Apple tech-support hotline. It's 800-275-2273 (that's 800-APL-CARE). Tech help is free for 90 days after you've installed Big Sur, or three years if you've bought AppleCare (Apple's extended-warranty program) for your Mac.

CHAPTER SIXTEEN

Eight Cool Features You Didn't Know You Had

You're probably too young to remember the first time a cool Mac feature blew people away. It was the day Apple co-founder Steve Jobs demonstrated the very first Mac by *pasting a picture into a word-processing document.*

It was 1984. The audience lost its mind.

Ever since then, Apple engineers have prided themselves on adding features that are powerful, useful, and—when possible—also *really cool.* As in magical.

These features often combine the real world and the digital one in wireless, wizardy ways: voice control of your Mac, projecting your Mac on your TV wirelessly, adding your written signature to a digital contract, and so on.

Siri

Siri is Apple's voice assistant, the one that began life on the iPhone in 2011 and is now built into every machine Apple makes—including, of course, the Mac.

Siri understands the same kinds of queries on the Mac that she does on the iPhone or iPad. You can ask her all kinds of questions, without ever worrying about the phrasing. If you want to know what the weather's going to be, you can say, "What's the weather outlook?" "Will I need a sweater tomorrow?" "Give me the Boston weather," "Will I need an umbrella in Cleveland?" or just about any variation. She answers you with a spoken voice, typed text, and often graphic images.

You can do more than ask her for information, though; you can also ask her to *do* things. You can say "Remind me about the muffins at 7:30 p.m.," or "Make an appointment with Chris on Monday at noon," or "Open Safari." She can even find files for you: "Show me the last email I sent to Alex," "Find all my pictures from the Grand Canyon last year," or "Find the spreadsheets I worked on last week."

If you've set up home-automation gear that bears Apple's HomeKit logo, you can even control your house: "Turn off all the lights," "Close the curtains," "Who's at the front door?" and so on.

Two Ways to Ask Siri

Not everybody wants Siri. That's why she comes turned off. To give her a try, you first have to turn on her master switch in **System Preferences→Siri→ Enable Ask Siri**.

Even then, Siri isn't listening all the time. For that, you'd need a human assistant, which costs a lot more.

Instead, she listens for a command or a question only when you do one of these things:

- **Click the Siri icon** on the menu bar.

- **Hold down ⌘-space bar** until the Siri panel appears.

Siri listens when you click the icon.   *Siri wins every trivia contest.*

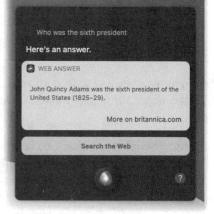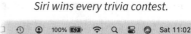

Siri in action

- **On the Touch Bar,** if your Mac has one (page 73), touch the Siri button.

On Macs made in 2018 or later, there's also a hands-free method: Just say, "Hey Siri." (This feature, too, requires setup. In **System Preferences→Siri,** turn on **Listen for "Hey Siri."** You'll be asked to speak a few sample Siri queries to teach her what you sound like.)

> **NOTE:** In **System Preferences→Siri,** you can change Siri's voice, gender, regional accent, and keystroke. Here, too, you can delete your dictation history, so Apple's computers no longer have any recordings of you issuing commands.

Any of these techniques opens the Siri palette in the upper-right corner of your screen. At this point, you can speak your question or command.

When you're finished with Siri, you can make her panel slide away by pressing the Esc key, clicking the ⊗ in the upper-left corner, or using a Siri command like "Goodbye," "See you later," or "Go away."

> **TIP:** If you're in a public library, it may be more polite to *type* your Siri queries instead of speaking them. To set this up, open **System Preferences→Accessibility→Siri.** Turn on **Enable Type to Siri.** From now on, the Siri panel contains a **Type to Siri** box. Type your query and then hit Return. (You can still speak requests.)

Questions to Ask

Siri is one with the internet; in fact, she can't do much without it. But with it, she's ready to help you with answers to all kinds of questions:

- **Search the web.** "Search the web for a used Tesla Model 3." "Search for banana bread recipes." "Search for news about the next *Die Hard* sequel." "Search the web for pictures of pineapples."

"What movie won the Best Picture Oscar in 1999?" "What's the tallest mountain in the world?" "When was Barack Obama born?" "How many days until Thanksgiving?" "How many inches in a mile?" "What's the exchange rate between dollars and yen?" "What's the capital of Germany?" "How many calories are in an Almond Joy bar?" "What's a 20 percent tip on 71 dollars for three people?" "What's the definition of 'erudite'?" "How much is 142 dollars in francs?"

"How old is Robert Downey Junior?" "When's the next solar eclipse?" "Give me a random number." "Graph x equals 2y plus 7." "What flights are overhead?" "Where was Gwyneth Paltrow born?" "How much caffeine is in a cup of tea?" "How much fat is in a Big Mac?"

"Search Wikipedia for John Lennon." "Tell me about Fiorello La Guardia." "Show me the Wikipedia page about bioluminescence."

- **Consult the clock.** "Time." "What's the time?" "What time is it in Dallas?" "What's today's date?" "What's the date a week from Friday?"

- **Find restaurants.** "Good Indian restaurants around here." "Find a good pizza joint in Boston." "What are the reviews for Red Lobster in Cincinnati?"

- **Ask for directions.** "Give me directions to the airport." "How do I get to 200 West 70th Street, New York City?" "Take me home." "Navigate to the nearest gas station."

- **Consult the Yellow Pages.** "Find pizza near me." "Where's the closest CVS?" "Find a hospital in Houston." "Search for gas stations." "What are the hours of the Apple Store?"

- **Check the weather.** "What's the weather today?" "What's the forecast for next week?" "What's the temperature in Austin this weekend?" "Check the forecast for Prescott, Arizona, on Friday." "Tell me the windchill in Juneau." "What's the humidity right now?" "How's the weather looking in Paris?" "How windy is it?" "Should I wear a jacket?"

"When's the moonrise?" "When will the sun set today?" "When will Mars rise tomorrow?"

- **Ask about stocks.** "What's Nike's stock price?" "What did GM close at today?" "How's the S&P 500 doing?" "How are the markets doing?" "What's Verizon's P/E ratio?" "What's Best Buy's average volume?"

- **Find out about your Mac.** "How much free space do I have?" "What's this Mac's serial number?" "What version of macOS do I have?" "How fast is my Mac?" "Is Bluetooth on?" How much iCloud storage do I have left?" "How much memory does my Mac have?"

- **Request movie trivia.** "Who was the star of *101 Dalmatians*?" "Who directed *Die Hard*?" "What is *Three Amigos* rated?" "What movies are opening this week?" "What's playing at the Loews 60th Street Cineplex?" "Give me the reviews for *Tenet*." "What are today's showtimes for *Star Wars XVIII: The Rise of the Dawn of the Force*?"

- **Request sports stats.** "How did the Dolphins do last night?" "What was the score of the last Marlins game?" "When's the next Cavs game?" "Are there any baseball games on today?" "Who has the best batting average in the American League?" "Who has scored the most goals in German

Movie and sports stats

soccer?" "Show me the roster for the Oilers." "Who is pitching for San Francisco this season?" "Is anyone on the Braves injured right now?"

- **Consult your address book.** "What's Ann's work number?" "Give me John Jacoby's office phone." "Show Hannah's home email address." "What's Sarah Cooper's home address?" "When is my wife's birthday?" "Show Avery Smalling." "Who is Payton Phoenix?" "Show my boss's work number." "Give me directions to my girlfriend's house."

> **TIP:** If you request information on people with relationships to you—mother, father, parent, brother, sister, son, daughter, child, friend, spouse, partner, assistant, manager, wife, husband, fiancé, boss—Siri can't comply until she knows who that is. So the first time you mention your brother (for example) in a query, Siri asks you for his name and then asks if you'd like her to remember that for next time. That's how she learns.

If, in speaking, Siri mispronounces somebody's name, say: "Learn to pronounce Siobhan Hbzrazny's name" (or whatever it is). With tremendous deference, Siri invites you to speak the correct pronunciation and then displays a set of ▶ buttons so you can pick the closest match from her offerings.

- **Look up your passwords.** "What's my Netflix password?" "Look up my Amazon password." "I need my Bank of America password." (She requires your Mac password or Touch ID before displaying your Safari Passwords screen.)

> **NOTE:** This feature is awesome! Admit it: You never knew it existed.

Commands to Issue

The only hard part about using Siri is *remembering to use her*—which entails remembering all the kinds of things she can understand. Now we're not talking about questions you can ask; we're moving on to commands you can give her.

She does come with a handy cheat sheet of commands; it appears when you click the ⑦ button on the Siri panel. You can also bring up that cheat sheet by asking, "What can I say?" "What can you do?" or "Help me!"

The Siri cheat sheet

Or you can retreat to the more complete cheat sheet that begins here:

- **Open apps.** "Open Notes." "Play Chess." "Launch Excel."

- **Change settings.** "Turn on Do Not Disturb." "Make the screen brighter." "Dim the screen." "Turn on Bluetooth." "Turn off Wi-Fi." "Turn on the screen saver." "Make the volume softer." "Mute the sound." "Louder."

- **Open System Preferences.** "Open Wi-Fi settings." "Open Displays in System Preferences." "Open Sounds settings." "Open wallpaper settings."

- **Go to folders.** "Open Documents." "Open the Trash." "Open the Applications folder."

- **Stop working.** "Put the Mac to sleep." "Shut down the Mac."

- **Find photos.** "Show me the Grand Canyon album." "Show me the videos from Christmas last year." "Get me the videos from New York." "Show me the slow-mo videos from Middlebury." "Give me the pictures from last summer." "Show me my cat photos." "Show pictures of Tia last year in Arizona."

- **Find files.** "Find the PDFs I worked on last week." "Show me the apps in my Downloads folder." "What was I working on yesterday?" "Show me the Microsoft Excel documents I worked on two days ago." "Show me the files I shared with Alex yesterday." "Round up all my Keynote files." "Where are the files tagged In Progress?" "List all the files with the word 'marina.'" "Let's see the files on my desktop." "Get my Vacation folder." "Show me the file I sent Stacy."

  You can further refine one of those requests. If you've just said, "Let's see the files on my desktop," you can then add, "Just the ones I edited this week."

- **Place calls.** "Call Janet." "Call the office." "Start a FaceTime call with Alton DeVries." "FaceTime Todd."

- **Listen to your email.** "Read my latest email." "Read my new email." "Any new mail from Blake today?" "Show new mail about the office party." "Show yesterday's email from Taylor."

- **Dictate mail.** "Email Charlie about the reunion." "Email my dad about the dance on Friday." "New email to Emery Holstein." "Mail Mom about Saturday's flight." "Email Richard Panton and Cynthia Powell about the picnic." "Reply 'Dear Finley (comma), thank you for your concern (period). My goldfish is just fine (period).'" "Email Carter and Tatum about their work on the Jenkins project and say, 'I couldn't have done it without you.'"

- **Check your calendar.** "What's on my calendar today?" "What's my day tomorrow?" "When's my next appointment?" "What's the rest of my day look like?" "What's on my calendar for April 4?" "When is my meeting with Rowan?" "Where is my next meeting?"

- **Make calendar appointments.** "Make an appointment with Jeri for Tuesday at 3 p.m." "Set up a haircut at 3." "Set up a meeting with Harold this Monday at noon." "Meet Bobby Cooper at eight." "New appointment with Dexter, next Wednesday at 7." "Schedule a Zoom call at 7:30 tonight in my office."

- **Reschedule appointments.** "Reschedule my meeting with Harold to a week from Monday at noon." "Move my 3 o'clock meeting to 3:30." "Add Rowan to my meeting with Jeri." "Cancel the Zoom call on Sunday."

- **Listen and reply to your texts.** "Read my new messages." "Read that again." "Read my last message from Frankie." "Reply, 'Oh wow (period). Why didn't I see that coming (question mark)?'" "Tell him I won't be there because I have to rearrange my sock drawer."

- **Send texts.** "Send a text to Eden London." "Send a message to Lennox saying, 'You're so fired.'" "Tell Noel I'm running late." "Send a message to Drew's mobile asking her to pick me up at the bus station." "Send a text message to (800) 922-0200." "Text Rowan and Jeri: Did you pick up the pizza?"

- **Operate your Notes app.** "Make a note that the back door key is under the mat." "Note: Do not put as much jalapeño into the recipe as it calls for." "Create a 'Things to do this weekend' note." "Find my Master Passwords note." "Show all my notes." "Start a note called 'Good TV.'" "Add *Breaking Bad* to my Good TV note."

- **Set a reminder.** "Remind me about Reese's birthday on August 25." "Remind me to bring the muffins to the meeting." "Remind me to take my hormone pill tomorrow at 7 a.m." "Set a reminder to change the air filter when I get to Mom's house." (That one is handy if you have an iPhone, because it knows your location. When you do, in fact, get near your mom's house, the reminder will pop up.)

- **Play music or podcasts.** "Play some Beatles." "Play 'Bohemian Rhapsody.'" "Play some blues." "Play my workout playlist." "Shuffle my 'Romantical Evening' playlist." "Play." "Skip." "Play podcasts." "Jump back 15 seconds." "Skip ahead four minutes." "Find the *Unsung Science* podcast in the Podcasts app."

If you're an Apple Music subscriber, you can *really* go nuts. Just ask for any song, album, or performer. "Play 'Rocket Man.'" "Show me some B.B. King albums." "Play 'Eleanor Rigby' next" (or "... after this song"). "Shuffle Billy Joel." "Play more like this." "Skip this song." "Play 'You and I' by Jacob Collier." "Play that song from the movie *Coco*." "Play the top song from 2007." "Play the top 20 songs of the 1980s."

And in midsong: "What song is this?" "What album is this from?" "Who's singing this?" "Like this song." "Rate this song two stars."

Siri's Personality

Believe it or not, Apple actually employs professional comedy writers whose sole job is writing wisecracks for Siri. (She'll never answer you sillily unless you start it by *asking* a silly question.)

For example, she's got a couple of dozen different answers for "What is the meaning of life?" (Example: "I Kant answer that. Ha ha.")

The web teems with collections of Siri's wit and wisdom. But for a small sampling, ask her:

"Are you married?"

"How old are you?"

"Can I borrow some money?"

"Do you like Android phones?"

"How much wood would a woodchuck chuck if a woodchuck could chuck wood?"

If the conversation stalls, try, "Tell me a joke." Or "Read me a haiku." Or "Tell me a story."

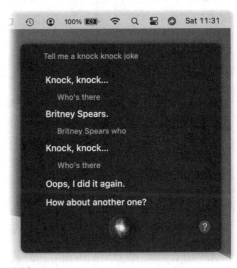

Siri humor

Two Ways to Dictate

As an interactive, voice-operated assistant, the Mac can both dish it out *and* take it. That is, you can speak to type. The Mac takes dictation.

Dictation can, in theory, be faster than typing, but take into account the time it will take to fix the mistakes the Mac makes.

Oddly, the Mac has *two different* dictation features. Dictation is the old one; Voice Control is newer. Each has its pros and cons:

Dictation	Voice Control
Types what you say	Types what you say *and* lets you control the mouse, keyboard, menus, and the rest of your Mac
You have to correct errors manually	You can correct errors by voice
Requires an internet connection	Does not require an internet connection
Maximum utterance length: 30 seconds at a time	No length limit
Types in any app	Doesn't work in some apps (like Adobe apps)

The actual act of dictation, and the resulting accuracy, is the same for both technologies.

Beginning Dictation

To turn on the dictation feature you want, begin in System Preferences:

- **Voice Control.** Open **System Preferences→Accessibility→Voice Control**, and turn on **Enable Voice Control**. If you have more than one microphone, choose the one you want to use.

> **TIP:** You'll get better accuracy if you're wearing a headset instead of relying on the Mac's built-in microphone.

- **Dictation.** Open **System Preferences→Keyboard→Dictation**. Click **On**.

Ready to try it out? Open an app that accepts typed text, like Mail, Notes, or TextEdit. Tell the Mac that you want to dictate like this:

- **Voice Control.** If the floating microphone palette says **Sleep**, then the Mac is already listening. (If not, click **Wake Up**.)

- **Dictation.** Press the fn key twice. (You can choose a different keystroke in **System Preferences→Keyboard→Dictation**.)

You can now begin speaking! Just talk normally; no need to speak loudly or jerkily. Don't forget, though, to speak the punctuation you want: "This is a

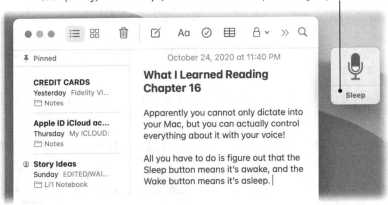

Click (or say, "Go to sleep") to make the Mac stop listening to you.

Voice Control microphone palette

test (dash)—of the Mac dictation system (period). I really (comma), really hope it works (period)." The transcribed text appears all at once.

You can also say "New line" or "New paragraph," "Open paren" and "Close paren" (when you want parentheses), "Quote" and "Unquote," "Smiley face" or "Frowny face," and just about any punctuation symbol invented: "percent sign," "at sign," "dollar sign," "copyright sign," "pound sign," and so on.

The Mac is smart enough to capitalize the first word of a sentence, but you can also say "Cap" before speaking any other word you'd like capitalized.

You can also say "Caps on" to make the Mac Capitalize Every Word Until You Say, "Caps off." Or say "All caps on" TO WRITE EVERYTHING IN CAPS—UNTIL YOU SAY "All caps off."

To make the Mac *stop* transcribing what you say, do this:

- **Voice Control.** Say "Go to sleep," or click **Sleep** on the microphone panel.

- **Regular Dictation.** Press the fn key twice.

The accuracy is OK, but you'll discover very quickly that it's not flawless. (Sometimes, you can't really blame the Mac. How is it supposed to know if you said *text wrap* or *tech strap,* or *movie clips* instead of *move eclipse*?)

To fix mis-transcriptions made with Dictation, use the mouse and keyboard, exactly as though you'd made typos yourself.

If you're using Voice Control, though, you have another option: correcting mistakes by voice.

Editing with Voice Control

Here's where Voice Control really shows off: It lets you edit text hands-free. Maybe you're correcting mistakes it made, or maybe you've just decided to *rewrite* something. Either way, while the Mac is listening, you can say things like these:

- **"Delete that"** or **"Scratch that"** to erase the last word or phrase you said. "Undo that" is pretty great, too.

- **"Move after 'Dear Mom'"** [or whatever] to move the insertion point to a certain point in the text.

- **"Move to end"** or **"Move to beginning"** of whatever you've been writing.

HOW TO CLICK ANYWHERE WITH VOICE CONTROL

Telling Voice Control to "Click OK" or "Click Send" works great—when there's a clearly labeled button to click. But what if you want to click a spot that's not labeled? What if you're trying to click an unlabeled toolbar icon? Or if you're trying to click a particular spot in Maps?

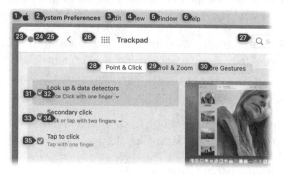

Apple came up with a couple of ingenious workarounds.

If the problem is that there's some icon or button that doesn't have a name onscreen, say, "Show numbers." The Mac superimposes tiny number tags on every clickable item in the window. Now you can just say, for example, "nine" to click whatever is identified by the 9 tag. The number tags appear on menu commands, too, which can make choosing from menus speedier.

And if you're trying to click a spot with no name *and* no number tag, you can say, "Show grid." This time, the Mac overlays a grid of numbered squares on the entire screen. ("Show *window* grid" confines the grid to the active window for better precision.)

Each time you speak the number of a grid square, the grid squishes down into just that square, making your number-calling increasingly more precise. Eventually, you'll have shrunken the grid enough that you can say, "Click seven" (or whatever the square's number is), even if it's as small as a single pixel. You can even say, "Drag two to nine" to drag from one spot to another. When you're finished with the gridding, say, "Hide grid." ⭐

- **"Move forward (**or **backward)"** and specify any number of lines, sentences, or paragraphs.

- **"Select all"** highlights all the text in the document. But you can also highlight only a single word or phrase ("Select 'Regarding last week's meeting'").

- **"Select next (**or **previous) word"** or character, sentence, line, or paragraph. Or specify how many: "Select next four words." You can also "Deselect that."

- **"Bold (**or **italicize,** or **capitalize) that"** to format text you've selected. Or specify the specific words you want to format.

- **"Correct 'oxymoron'"** (or **"Correct that").** Voice Control shows you a list of alternative interpretations, numbered for your convenience.

Click, or say, "Two."

Alternative words

- **"Replace 'you had the gall to' with 'you chose to.'"** That right there is an amazing feature: You can speak to replace any word or phrase, without fiddling around with commands to move the selection first.

All of this works in Apple apps—and Microsoft's, too.

Control the Mac by Voice

But dictation is only half of Voice Control's purpose in life. It also lets you control *the entire Mac* by voice: opening apps, clicking buttons, changing settings, and so on. Apple designed it for people who can't use a mouse, keyboard, or trackpad, of course, but you may find situations where it's handy no matter who you are.

Here are some of the most useful verbs you can use:

- **"Click."** You can say, for example, "Click OK." "Click Reply." "Click Done." "Click and hold mouse." "Double-click." "Click File menu." "Triple-click."

- **"Open"** and **"Close."** "Open Finder." "Close Siri." "Launch Photos." "Save document." "Quit Maps."

- **"Scroll"** and **"Zoom."** "Scroll down." "Zoom out."

To see the complete list of Voice Control commands, open **System Preferences→Accessibility→Voice Control→Commands**. (You can turn off individual commands here, too. That's useful if you find yourself triggering a command accidentally.)

Or say, "What can I say?" to see Voice Control's master command list.

Voice Control cheat sheet

Voice Control is smart enough to ignore any talking it hears that *isn't* one of the commands it knows. If you say "Open Notes," Notes opens. But if you say, "Where were you last night?" Voice Control sits tight and shuts up.

TIP: In **System Preferences→Accessibility→Voice Control→Commands**, you can define new commands to replace ones that strike you as clumsy or too long. In the **Perform** pop-up menu, you'll see that your custom commands can open things in the Finder, open websites, paste text or data, press any keyboard combination, choose from a menu, or run a workflow.

A workflow, since you asked, is a canned sequence of steps you set up in Automator, the Mac's app for creating very simple programs. Using this feature, some Googling, and a good deal of patience, you can make Voice Control do things Apple never even dreamed of.

The Mac Reads to You

Believe it or not, the Mac can read anything on the screen aloud. An email, a web article, a note, an ebook—anything.

Most Apple apps—Safari, Notes, Pages, Mail, TextEdit, and so on—have a built-in speaking command. When you find some text worth hearing, right-click or two-finger click in the window; from the pop-up menu, choose **Speech→Start Speaking**.

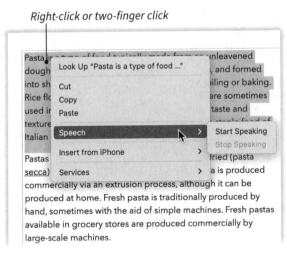

Right-click or two-finger click

Start Speaking command

But macOS goes much further than that. It can read any text in any app. You can choose from a long list of synthesized voices for your Mac; some of them sound stunningly human.

To get started, open **System Preferences→Accessibility→Spoken Content**.

Use the **System Voice** controls to choose a speaking voice for your Mac. Because voices take up a lot of disk space, you have to download most of them using the **Customize** command. Alex, Samantha, and Tom are particularly lifelike. But you can also choose English-speaking voices with British, South African, Irish, and Australian accents. Some of the voices in the **English (United States)—Novelty** category actually *sing* to you.

*Choose one of the seven starter voices...or choose **Customize** for a whole catalog.*

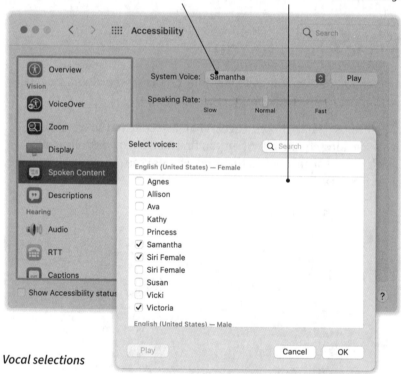

Vocal selections

Then turn on **Speak selection**. This is the feature that lets the Mac read *any* selected text, in any app. Press Option-Esc to begin the reading, and press again to stop it.

> **TIP:** Click **Options** to change the keyboard shortcut that makes the speaking start or to specify whether, and how, the words are highlighted on the screen as the Mac reads to you.

Your Mac on TV

You can probably think of all kinds of reasons you might want to project your Mac onto a big TV. It's the perfect way to show slideshows and movies to a group sitting around the room. It's super-handy for making presentations from Keynote or PowerPoint. It's great for teaching. It's essential for playing games, especially when friends or family want to watch.

You can connect your Mac to a TV either with a wire or without. Both require some extra hardware:

- **Connect a wire.** Most TVs made since about 2003 have an input jack called HDMI. Every Mac has a video-output jack—on recent models, it's a Thunderbolt 3 (aka USB-C) jack, although the types have varied over the years. You just need to buy the right cable. Once you connect it to the TV, it acts as though it's a second monitor, and page 28 takes it from here.

An HDMI connector

- **Connect wirelessly (with an Apple TV).** The Apple TV is a little black box that brings hundreds of internet "channels" (including your Netflix, Hulu, HBO, Amazon Prime, and other online subscriptions) to the TV. But it has a stealth feature, too.

It's called AirPlay, and it's Apple's wireless video-sending technology. When it's time to present your Mac's screen, open the Control Center (🔛) and click **Screen Mirroring**. Choose the name of your Apple TV (you may also see the name of your TV or receiver if it works with AirPlay), and presto: Your TV lights up with the Mac's image.

> **NOTE:** The very first time you try this, a huge four-digit number appears on the TV for you to type into the Mac. That's a security thing, so your next-door neighbor can't surprise you by projecting raunchy videos onto your TV when you're trying to watch *60 Minutes*.

At this point, you can choose how you want the TV to act: as a duplicate of your Mac screen or as an extension of it. It starts out mirroring, because that's almost always what people want. But if you're not most people, you can use the 🖵 menulet to choose **Use As Separate Display**. (If you don't see the 🖵 menulet, open **System Preferences→Display** and turn on **Show mirroring options in the menu bar when available**.)

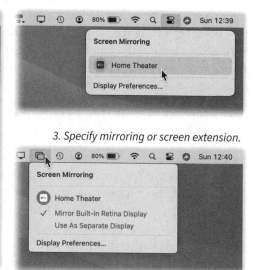

1. In the Control Center, click Screen Mirroring.

2. Choose your Apple TV (or AirPlay-equipped TV).

3. Specify mirroring or screen extension.

AirPlay to your TV

To finish with your presenting, click the ⛃ and click **Screen Mirroring** to turn it off. Wasn't that cool?

Make a PDF

You're probably already aware of, and maybe even grateful for, the ubiquity of PDF documents. If not, you can read page 296 as a refresher. How lucky we are to live in a world with a universal format for documents—fonts, graphics, layouts—that looks identical on Mac, Windows, phone, tablet, or the web!

What you may be less aware of, and therefore less grateful for, is that you can turn *any Mac document* into a PDF file. If you can print it, you can PDF it. That's a huge, huge feature. It means no matter what weirdo app you're using that very few other people own—a sheet-music app, a garden-plot diagramming app, a blueprint app—you'll be able to distribute your work to the world.

The trick is to start out like you're going to print the document. In most apps, that's **File→Print**.

And right there, at the bottom of the Print box, is your PDF pop-up menu. Choose **Save as PDF**, specify a name and folder for the PDF file that's about to be born, and hit **Save**. That's all there is to it: no special software (or special computer skills) required.

		Printer:	HP LaserJet Professional P1...	
		Presets:	Default Settings	
		Copies:	1	

1 of 1

Does Your Carburetor Need Cleaning?
Alex Arsenal, chief tech, Carparts 'n Carports

☑ Two-Sided

Pages: ○ All
⦿ From: 1   to: 1

? | Show Details | PDF ⌄ | | Cancel | Print

Open in Preview
Save as PDF
Save as PostScript

Send in Mail
Save to iCloud Drive
Save to Web Receipts

Edit Menu...

Creating a PDF

Capture the Screen

Why would you ever want to capture a picture of your screen? Maybe because you're trying to show somebody in tech support the weird glitch you're seeing. Maybe because you want to tweet about something funny you see onscreen. Maybe because you're writing a book about macOS Big Sur.

All you have to remember is the not-very-mnemonic keyboard shortcut Shift-⌘-5. It makes the screenshot toolbar appear at the bottom of the screen (although you're welcome to drag it anywhere).

Why is a whole toolbar necessary for making a screenshot? Because you need options! Maybe you want to capture the whole screen, or maybe just a part of it. Maybe you want the arrow cursor in the shot, and maybe you don't. Maybe you want a still image, or maybe you want to record a *video* of your activity.

Here's how to run it:

- **Capture a still image.** If you'll need a five- or ten-second countdown to get the screen set up the way you want it, use the **Options** pop-up menu to choose a timer. While you're in Options, specify where you want the screenshot to go once you have it: to the desktop, for example, or into one of your graphics apps, or directly to Mail as an attachment.

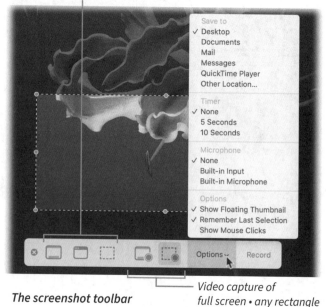

Still capture of
full screen • window • any rectangle

Save to
✓ Desktop
Documents
Mail
Messages
QuickTime Player
Other Location...

Timer
✓ None
5 Seconds
10 Seconds

Microphone
✓ None
Built-in Input
Built-in Microphone

Options
✓ Show Floating Thumbnail
✓ Remember Last Selection
Show Mouse Clicks

The screenshot toolbar

Video capture of
full screen • any rectangle

Now click one of the still-capture buttons: **Entire Screen**, **Selected Window** (neatly isolates whichever window you click), or **Selected Portion** (a chunk of screen you isolate by dragging the corners of the dotted-line rectangle, then press Return or click **Capture**).

- **Capture a screen video.** Once again, use the **Options** menu to set up a timer, if you wish, and to specify where the saved video winds up.

  If you turn on **Show Mouse Clicks**, in the finished video, a little circle will appear around the cursor each time you click. That's helpful for viewers to figure out where and when you were actually clicking.

  Now click **Record Entire Screen** (to capture the big picture) or **Record Selected Portion** (and then drag to isolate the rectangular area you want to record). Finally, click **Record** on the toolbar.

  At this point, the Mac is "filming" everything you do.

NOTE: Sadly, the Mac doesn't capture the Mac's sounds—only the video. (Commercial apps like ScreenFlow can record both sound and picture.) On the other hand, you can narrate what you're doing; that is, the Mac can incorporate sound from your microphone into the screen video. Just choose your microphone's name from the **Options** menu before you start.

To end the recording, click ⊙ in your menu bar.

After you take a shot or a video, a thumbnail of your new capture appears at the lower-right corner of your screen. If you click it, it opens up for editing, as described next. If you ignore it for six seconds, the thumbnail slides away, and your screenshot becomes an icon on your desktop (or wherever you directed the Mac to save it).

> **NOTE:** If that six-second thumbnail business seems unnecessary, open the **Options** menu and switch off **Show Floating Thumbnail**.

Editing Your Shot

If you click the thumbnail in time, it opens into a special floating editing mode. If you took a still screenshot, you get the standard macOS markup tools, which are described on page 297, plus **Rotate** and **Crop** buttons. Here's your chance to annotate the screenshot, draw arrows, circle important things, and so on.

TWO MORE WAYS TO GET SCREENSHOTS

Years before Apple invented the Shift-⌘-5 toolbar, it provided two far simpler screenshot tools. They're still available when you don't need all the hullabaloo about videos, countdowns, choice of folder location, and so on.

For a full-screen shot, press Shift-⌘-3 to create a picture file. To choose a rectangular portion of the screen, it's Shift-⌘-4 (and then drag to select the area you want).

The Shift-⌘-4 variant, by the way, harbors a whole host of handy special options; most are designed to help you isolate exactly the right target screen area. Once you've begun dragging to isolate your rectangle, you can press the space bar to freeze your selection shape; now you can move it, by moving the cursor, to a new position. The Shift

key confines your dragging to either horizontal or vertical motion. The Option key makes your rectangular area grow from the center instead of the diagonal corners.

Here's a cool trick, too: Once you've pressed Shift-⌘-4, you can neatly snip out just one window or open menu by tapping the space bar (instead of dragging a rectangle). Click the window you want.

If you're pressing the Control key as you release the mouse, by the way, you copy the screenshot to your Clipboard, ready for pasting, instead of saving it as a file on your desktop.

Oh—and if you decide to cancel this screenshot attempt altogether, press the Esc key. ✦

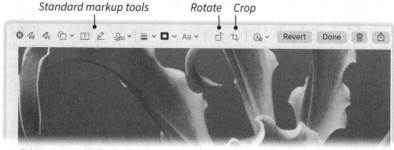

Standard markup tools   *Rotate*   *Crop*

Editing a screenshot

Click **Revert** to undo all the edits you've made, **Done** to save your work as a file on the desktop, 🗑 to delete the shot, or ↥ to send your screenshot to someone else.

If it's a screen video, the editing mode gives you the chance to trim the dead air off the beginning or the end (as described on page 300).

Screen Time: Protect Your Kids

On top of everything else we have to worry about, now there's the screen-time thing. We worry that too much time spent using electronics is doing long-term damage to our kids' brains—and maybe even our own.

Even Apple is worried about it. That's why it came up with Screen Time, a feature that monitors what apps you're using, and for how much time, and on which Apple gadgets (Mac, iPhone, iPad). You can also set daily limits for each category—two hours a day of social media, for example—for you or for your kids.

Every Sunday, a notification from Screen Time appears on your Mac screen, offering a weekly report. It's a handsomely designed summary, full of graphs that show how much time you and your other family members spent staring at their screens, what apps they were using, and how many times they exceeded the limits you set.

Turning on Screen Time

To get started, open **System Preferences→Screen Time**. Click **Options** and then **Turn On**.

Screen Time can, if you wish, add up the time you spend hunched in front of *all* your Apple machinery: the Mac, the phone, the tablet. Turn on **Share across devices** if you want that kind of data horror show. (You must turn that switch on separately on each device you use.)

Setting Up a Passcode

Of course, Screen Time won't do you or your kids much good if they can just turn it off and play 11 straight hours of Minecraft. Enter: the **Use Screen Time Passcode** checkbox in **System Preferences→Screen Time→Options**.

Choose your kid's name from the top-left menu. The software will balk if you've given the kid an administrator account (page 317) instead of a standard account. What good would a passcode do if the kid could just change it—or delete you?

Now turn on **Use Screen Time Passcode**. You're asked for a four-digit number. Without it, nobody can turn off Screen Time or change its settings.

[Names Menu]

This pop-up menu lists your family members' names, assuming you've turned on Family Sharing. Use it to specify whose screen time you want to disapprove of.

App Usage, Notifications, Pickups

These three tabs reveal graphs that illustrate how much time the chosen family member has spent on Apple apps this week; how many notifications bombarded her; and how many times she woke and unlocked her gadgets.

For each morsel, you can limit the graph by date or by Apple device.

When you're viewing App Usage, the bottom half of the screen shows how much time this person has spent in each of the **Apps**—or each of the **Categories**, like **Social** networks, **Productivity & Finance**, **Creativity**, and so on.

This week *Past 24 hours* *Limits and Info buttons*

Setting up Screen Time

> **TIP:** And how do you know what category the Mac considers each app to be in? Point to the **Limits** column without clicking, and click the ⓘ button.
>
> And while you're pointing without clicking: You can also click the ⌛ icon right there to set up a daily time limit for that app or category, as described momentarily.

Downtime

Downtime means "enforced no-gadgets time." For the family member you've selected, you invoke it by clicking **Downtime** and then **Turn On**. Use **Every Day** to establish the no-gadgets hours—11 p.m. to 8 a.m., maybe. Or use **Custom** to establish separate schedules for each day, so weekends can be different from school nights.

When Downtime is five minutes away, a warning appears. Then, at the appointed time, the app windows all go white and their icons dim.

If you've set up Downtime for *yourself,* its onset is not much more than a slap on the wrist. You can click **Ignore Limit** and then click **One More Minute, Remind Me in 15 Minutes,** or **Ignore Limit For Today.** Or you could just turn Downtime off.

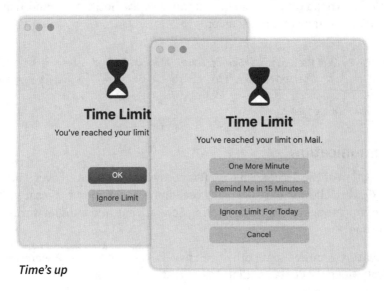

Time's up

But if you've added a passcode—for your kids, for example—there *is* no **Ignore** button. Lucky you—teachable moments every single day!

App Limits

You can also set daily time limits for individual apps or app categories.

To begin, click **App Limits,** and then **Turn On,** and then **+.** Turn on an app category to establish a daily time limit for that group of apps—or expand a category to see the apps in that category, to limit them one by one.

Dial up a maximum amount of time per day. Repeat with other apps and categories.

Click **Done.**

So what happens when your daily limit approaches for an app (or app category)? You get a five-minute warning. Here again, you can click **Ignore Limit**—*unless* you're blocked by a passcode *and* **Block at end of limit** was turned on for this app or category. Which probably means you're a kid.

In that case, the kid gets a new option: **Ask For More Time**. That button sends a request to you, the all-knowing parent, wherever you are in the world. If your kid has been very, very good, you can click the notification and, using its pop-up menu, approve an extension.

> **TIP:** In **System Preferences→Screen Time→Always Allowed**, you can designate certain apps this person is *always* allowed to use, even during Downtime. Maybe there are apps that you don't think should ever be off-limits—for example, a guitar-practice app, a jogging-coach app, or a meditation app.

Communication

This section of the Screen Time preferences gives you control over who your kids are allowed to reach using the Phone, FaceTime, or Messages. You can set up one set of conditions during Screen Time and another during Downtime.

This feature requires that both of you have turned on **System Preferences→Apple ID→iCloud→Contacts**. iCloud can't restrict your contacts if it can't see your address book.

If you click **Turn On** (lower right), you can choose from your own Contacts a list of people your kid is allowed to contact. Then you can limit communications to those people, or to groups that *include* at least one of those people.

And now, two items of not-so-fine print. First, your kid can always dial 911. Good.

Second, your kid can always bypass all this by using WhatsApp, WeChat, Facebook Messenger, or any other non-Apple app. Oh, well.

Content & Privacy

Here's where the Mac's parental controls live. You can use them to block stuff that might corrupt your kid's mind, like pornography and dirty words, as well as stuff that might corrupt your Visa card, like purchases of music, movies, and apps without your awareness.

In **System Preferences→Screen Time→Content & Privacy**, click **Turn On**.

You can set up restrictions in four categories:

- **Content** primarily means online pornography. **Limit Adult Websites** consults a blacklist that Apple maintains (known naughty sites); **Allowed Websites Only** is the opposite. It's a whitelist that permits *only* the wholesome sites listed here. You can click **Customize** to add new sites to the list.

  This tab also lets you block dirty words in the Dictionary app and offers limits to playing games in Game Center, Apple's game-coordination app.

- **Stores** can limit what **Movies, TV Shows**, and **Apps** your kids can consume from Apple's online stores; the pop-up menus specify ratings like **PG-13** and **R**. You can also use the **Explicit** checkboxes to block books, music, podcasts, news stories, and music videos that contain dirty words.

  Also tucked here: permission to install apps, delete apps, and make in-app purchases (buying new levels or weapons within a game, for example).

- **Apps** are apps and features that you may not want your kid to be allowed to use, on both the Mac and, in the lower list, iPhone or iPad.

- **Other** means "various other things you may not want your kids changing," like their iOS passcodes, email and social-media accounts, TV provider, and so on.

Overall, Screen Time offers a lot of controls that give you a lot of control. It can be dizzying, it should involve some open conversations with your kids, and it may involve a few days of debugging and troubleshooting. ("Hey, Parent—you wanna know why I didn't respond to your text? Because your stupid parental controls locked me out of Messages halfway through social studies!")

But if you're at all worried about your kid's screen time and exposure to the worst of the internet, at least Screen Time gives you a fighting chance.

Adapt the Mac

It would be hard to imagine a computer more generous with accessibility features. If you have trouble with seeing, hearing, or mobility, this is the machine you want.

The world headquarters for these accessibility features is **System Preferences→Accessibility**.

Help with Seeing

Some of the options on the **Display** tab make the Mac more usable no matter how good your vision is. In particular:

- **Reduce transparency,** on the **Display** tab, makes the background of the menu bar, menus, and panels *white* instead of translucent. Yeah, it might be cool that you can see your desktop wallpaper shining through your menus—but it can also make the menus harder to read.

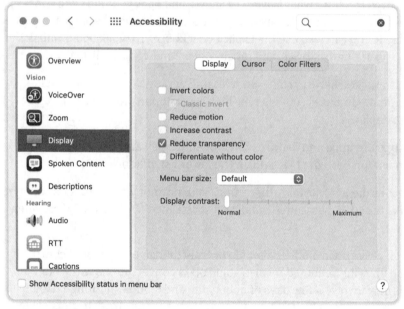

Accessibility preferences

- **Menu bar size,** also on the **Display** tab, makes the typeface of the menu bar and menus a *tiny* bit bigger—like a point or two bigger. You have to log out and log back in to see the difference.

- **Cursor size,** on the **Cursor** tab, is fantastic. As screen resolution gets higher these days, the pointer looks smaller and smaller. This slider lets you choose a more visible cursor.

> **TIP:** **Shake mouse pointer to locate** is fantastic, too. Use it when you can't see where the cursor *is*. Rapidly shake your mouse, or your fingers across the trackpad, to make the cursor suddenly get *huge* for a moment so you can spot it.

Fool around with the other options on the Display tab to see if they make things easier for you to see. If you're color-blind, for example, the **Color**

Filters shift *everything* on the screen to colors you can distinguish. If you have trouble reading, try **Invert colors**; it swaps the screen's colors like a film negative—black for white, red for green, blue for yellow—for better contrast. (It's smart enough not to affect photos, videos, or app icons.)

On the **Zoom** tab, you can set up temporary magnification of certain screen elements, as described on page 62.

And on the **VoiceOver** tab, you can turn on the Mac's screen reader, in which the Mac speaks aloud the name of everything under your cursor. It's possible, with practice, to operate the Mac without being able to see anything at all, thanks to VoiceOver and its many options. A full-blown user guide for VoiceOver awaits at apple.com/voiceover/info/guide/.

Help with Hearing

If you're deaf, partly deaf, or working with your speakers muted out of courtesy, you may appreciate **Flash the screen when an alert sound occurs**. It produces a white flash across the screen whenever the Mac would otherwise beep to get your attention.

Play stereo audio as mono is useful if you have hearing loss in one ear, or if you're wearing only one earbud. Now you won't miss out on half the sound in a stereo mix.

The **RTT** tab lets you set up a real-time text system, which permits deaf people to make phone calls by typing.

Help with Mobility

If you have limited motor control, Voice Control (page 384) should have gotten your attention. It's designed to let you operate everything on the Mac with voice commands.

The **Keyboard→Hardware** tab offers a range of accommodations, including **Sticky Keys** (lets you press multiple-key shortcuts like Shift-⌘-5 one key at a time) and **Slow Keys** (doesn't register a key press until you've held the key down for at least one second, to help screen out accidental keystrokes). The **Keyboard→Pointer Control** tab lets you adjust how close together two clicks must be for the Mac to consider them a double-click instead of two separate clicks.

And on the **Keyboard→Accessibility Keyboard** tab, Apple offers a mind-blowingly flexible feature for people who have trouble typing and just about anyone else. You can use it to build an onscreen keyboard or palette of any design, whose buttons type out canned bits of text, open apps, adjust Mac settings, and so on.

Copy and paste buttons into an arrangement you like.

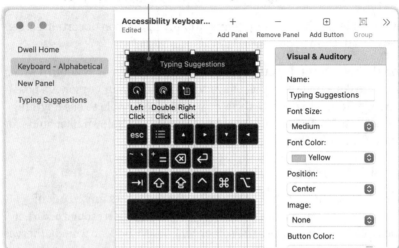

Accessibility Keyboard

The **Trackpad Options** and **Mouse Options** let you fine-tune your pointing-device behavior in strategic ways. You can, for example, adjust the tracking speed of your trackpad or mouse (how fast the cursor moves relative to your finger or mouse), and you can set up trackpad dragging, which means dragging objects without having to keep the trackpad clicked down.

In **Pointer Control→Alternate Control Methods**, you can turn on Mouse Keys, which lets you move and click the cursor using the number pad (on an extended keyboard) instead of the mouse or trackpad. You can hold down the 5 key to click, or move the cursor by pressing the eight directional keys around it.

> **NOTE:** In fact, these options even let you trigger cursor movements and mouse clicks using only your head turns and facial expressions, as viewed by the Mac's camera. In principle, you can use a Mac even if you have no use of your arms at all.

Help with Accessibility

There are a lot of people in the world, with a lot of different challenges. Clearly, Accessibility is a vast area—both the topic, and the software controls. Apple maintains an entire website dedicated to its assistance features: apple.com/accessibility.

Meanwhile, there are so many features to keep track of that macOS offers easy-access on/off switches, so you can summon them at will. In **System Preferences→Accessibility**, for example, you can turn on **Show Accessibility status in menu bar**; now all the features described here, and more, are only a quick click away.

> **TIP:** You can edit this menu. It lists only the features you've turned on in **System Preferences→Accessibility→Shortcuts**.

Or, if you don't need an entire menu using up space all day long, you can also summon a dashboard of Accessibility features and their on/off checkboxes. Just press Option-⌘-F5—or, on a Touch Bar laptop, fast-triple-click the Touch ID fingerprint reader instead. (These are audible click-downs, not just touches.) The options on this dashboard, too, reflect only the features you've turned on in **System Preferences→Accessibility→Shortcuts**.

Installing macOS Big Sur

I f you have a Mac running macOS Big Sur, then one of two things must have happened. Either the Mac entered your world with Big Sur already on it, or you upgraded to Big Sur from an earlier version of macOS. If one of those is the case, you can skip this appendix for now.

But if your Mac is running an older version of macOS—Catalina, say, or Mojave—then this appendix is for you. It's your guide to upgrading to Big Sur.

Horsepower Requirements

Like any major West Coast operating-system company, Apple does its best to ensure that each new OS release works on as many older computer models as possible. But sooner or later, the company has to draw a line. And that line says, "C'mon, guys, your Mac is now eight years old and too ancient to run our latest software. Get a new Mac, or quit complaining."

For Big Sur, here's what your Mac needs:

- **Relatively recent guts.** Big Sur can run on any MacBook Pro, MacBook Air, or Mac Pro made since 2013; any iMac or Mac mini made since 2014; or any MacBook (not "Air" or "Pro") made since 2015. In other words, a few 2012 and early 2013 Mac models that were still capable of running the previous macOS (Catalina) are no longer welcome to the club.

- **Some storage space.** You need at least 20 GB free on your Mac to upgrade to Big Sur. More is always better.

- **A lot of memory.** Apple recommends at least 4 GB of memory (RAM), but more makes the Mac run faster and smoother. (To check, choose →About This Mac.)

- **A backup.** MacOS upgrades rarely go wrong—maybe 1 in 10,000—but sometimes, the stars do misalign. That's why experts insist that you back up your entire Mac before you begin an OS upgrade. You can use Time Machine (page 64), for example, to do the job.

The Standard Installation

Upgrading to Big Sur doesn't touch anything that's already on your Mac—your files, folders, apps, and settings. The installer neatly works around them, so your stuff will be just as you left it, except it will live in a much nicer-looking world.

Most people, most of the time, don't have to seek out Big Sur; it comes to them. A notification bubble appears, letting you know that a software update is available. If you missed that, you can also open **System Preferences→ Software Update** and click **Update Now**.

> **NOTE:** If you're running an old version of macOS—High Sierra, for example—here's yet another way to get the upgrader: In the Mac App Store app (page 131), search for Big Sur, and then click **Get** and then **Download**. (It's a huge file—12 gigabytes—so be patient. Or take your Mac to an Apple Store to use its free and fast Wi-Fi.)

The Big Sur installer

When the download is over, the macOS installer opens automatically. Click **Continue**, then **Agree**, then **Agree** again. (Lawyers!)

Now you're supposed to choose a disk or partition of a disk where you want to install macOS. For most people, the main internal drive is the obvious choice, but if you have other drives, you can click **Show All Disks** to see their icons.

> **NOTE:** If you're installing onto a second drive, you may be offered a choice of drive-formatting schemes. In general, you want **Mac OS Extended (Journaled)**, which blesses you with a shorter startup time and better file safety if your Mac crashes.

Finally, you arrive at the **Install** button. Click it and enter an administrator's account name and password (page 317).

The installation usually takes 20 minutes or so, and it involves the Mac restarting a couple of times.

> **NOTE:** Because it's risky to run out of power during critical surgery like this, if you're on a laptop, the Mac suggests that you plug it in first.

After the final restart, you've successfully installed macOS 11.0 Big Sur—but you have to make a few more choices before you can dive in.

The Setup Assistant

The first app to run after a Big Sur installation is called Setup Assistant. It guides you through a bunch of different preference-setting sessions.

The options you see depend on whether you installed Big Sur onto a drive containing a previous macOS version or you're starting from scratch.

After Upgrading

If you've already been using your Mac for a while and you've upgraded it to Big Sur from a previous macOS version, then the Mac doesn't have many questions for you. It already knows most of your preferences.

Therefore, you can log in as usual. The Mac asks for your iCloud name and password (you can sign up for one now, in the unlikely event that you don't have an iCloud account). If you have two-factor authentication turned on (page 151), Apple now sends a confirmation to your other Apple gadgets; type the code in as requested.

Now successive screens may ask you for permission to send analytics to your apps' software companies (that is, after a crash, is it OK to send diagnostic log files?). You may be asked if you'd like to turn on dark mode (page 61), Siri (page 371), and other features.

Finally, you arrive at the desktop, in all of its gorgeous Big Sur–redesigned glory. Get surfing!

A New Mac (or an Empty Drive)

In two situations, the Setup Assistant asks a much longer list of questions. They are (a) you're working with a brand-new Mac or (b) you've installed Big Sur onto a completely empty drive.

On each screen, make your choice and then click **Continue**.

> **NOTE:** You'll notice that a large percentage of these screens inform you about certain kinds of data Apple intends to collect (anonymously, of course) and what it intends to do with that information. Each screen also gives you a chance to opt out of said data harvesting. This is all part of Apple's corporate insistence that you, and only you, should be in control of your data—and nobody else (*cough* Facebook and Google *cough*).

MAKE YOURSELF AN INSTALLER DISK

Back in the olden days—that is, until 2012—each new version of macOS came on a DVD, or even a pile of CDs. Installing those versions was slow and tedious, but at least you always had a backup installer.

Nowadays, of course, macOS updates always come from online—from The Cloud. That's handy, because you don't have any physical discs to lose. But it's also annoying when you want to install Big Sur on a whole classroom full of Macs, for example. Are you really going to download the 12-gigabyte installer 30 times?

Fortunately, you can turn a USB flash drive (16 GB or more) into an installer disk that bypasses the need to download the installer from Apple more than once.

There are a lot of steps involved, as well as a complicated Unix command you type or paste into Terminal. The web is full of excellent illustrated tutorials and videos, like this one: https://www.macworld.com/article/3566910/.

But when it's all over, you've got yourself a flash drive you can use to install macOS Big Sur on any compatible Mac. Here's how you do that:

Turn off the Mac. Plug in the flash drive. When you turn the Mac back on, hold down the Option key until a choice of startup disks appears. Click your flash drive; the Mac starts up from it; you log in.

Finally, you arrive at the Recovery screen shown on page 368. Click **Install macOS**, and off you go. ★

Here are the screens you may encounter:

- **Select Your Country or Region.** Click where you are. The next screen confirms your chosen languages and input source (keyboard layout, for example). Click **Continue** each time.

- **Accessibility.** Here's your chance to read about some of the assistance features described in the previous chapter and turn them on right at the beginning. Click **Vision, Motor, Hearing,** or **Cognitive** for a tour of the assistive features in each category and the option to turn on each one.

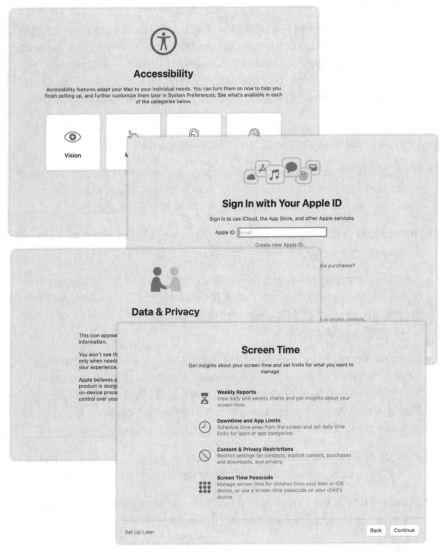

The Setup Assistant

- **Select Your Wi-Fi Network.** Makes sense that one of your very first acts with your new Mac should be helping it get on the internet. If there is no Wi-Fi hotspot nearby, click **Other Network Options** to specify, say, a wired or cellular network hookup.

- **Data & Privacy.** Here's a declaration about how seriously Apple takes your data privacy, with a link to an even longer declaration about it.

- **Migration Assistant.** This app can transfer everything from an old Mac (or even an old Windows PC) to your new Mac, including apps, files, folders, and settings.

  You can choose **From a Mac, Time Machine backup or Startup disk** (connect your old Mac, backup drive, or external drive to this one over a network, even a wireless one), and **From a Windows PC** (ditto). Click **Continue** if you actually want to do the data migration, or **Not Now** otherwise. (You can always run Migration Assistant later. It's in your **Applications→Utilities** folder.)

  Keep in mind that the data transfer can take a very long time.

- **Sign In with Your Apple ID.** Here's where you enter your Apple ID (your email address and iCloud password; see page 150). If you don't have one, you can either **Create new Apple ID** or click **Set Up Later**.

- **Terms and Conditions.** Apple's lawyers say hi. Click **Agree**, and then **Agree**. It may also help if you nod your head vigorously.

- **Create a Computer Account.** You're about to create the Mac account you'll use to log into this computer. It's an administrator account. You can choose an account photo, a password, and a password hint; all of this is described starting on page 316.

- **Express Setup** is a set of standard settings Apple assumes you don't want to argue with. If you click **Continue**, you're electing to turn on Siri, turn on Location Services (for use by Maps and other features), and share log data with Apple's programmers when your Mac crashes. If you prefer, you can click **Customize Settings** to deny these features one at a time.

- **Analytics** is where you can opt out of having log data sent automatically to software companies when their apps have crashed on your Mac.

- **Screen Time** is the time-logging and kid-tracking feature described on page 392. You can click **Continue** or (if you want to decline for now) **Set Up Later**.

- **Touch ID.** If your laptop has a fingerprint sensor, and if you have a fingerprint, you now get the chance to train the former to recognize the latter.

- **Siri.** Here's the master switch for the Siri feature (page 371)—and, if you find the whole idea of a digital assistant creepy, your chance to opt out.

  If you leave Siri on, the next screen is Improve Siri & Dictation. Are Apple quality-assurance staff allowed to store and maybe analyze your Siri and dictation utterances for the purposes of making these features work better? (They're completely anonymous, of course.) Click either **Share Audio Recordings** or **Not Now**.

- **Choose Your Look.** **Dark** or **Light**, O wise one? Or **Auto**, meaning that you'll get dark mode only at night? (See page 61.)

After your final **Continue** click, and a moment of Mac contemplation, you wind up at the macOS desktop, just as described in Chapter 4.

About the Author/Illustrator

David Pogue was the weekly tech columnist for *The New York Times* from 2000 to 2013. Today, he's a correspondent for *CBS Sunday Morning*, a *New York Times* bestselling author, host of 20 science specials on PBS's *NOVA*, and host of the podcast *Unsung Science with David Pogue*.

He's won five Emmy awards, given five TED talks, and written or co-written more than 120 books, including dozens in his *Missing Manual* series; six in the *For Dummies* line (including *Macs*, *Magic*, *Opera*, and *Classical Music*); two novels (one for middle-schoolers); and three books of essential tips and shortcuts: *Pogue's Basics: Tech*, *Pogue's Basics: Life*, and *Pogue's Basics: Money*.

In his other life, David is a magician, a funny public speaker and a former Broadway show conductor. He lives in Connecticut and San Francisco with his wife, Nicki, and a blended family of five fantastic kids.

You can find a complete list of his columns and videos, and sign up to get them by email, at authory.com/davidpogue. His website is davidpogue.com; on Twitter, he's @pogue. He welcomes feedback about his books by email at david@pogueman.com.

About the Creative Team

Julie Van Keuren (editing, indexing, interior book design) spent 14 years in print journalism before deciding to upend her life, move to Montana, and live the work-from-home dream before it was cool. She now provides skilled editing, writing, book layout, and indexing to a variety of terrific clients. She and her husband have two adult sons. Email: JulieVanK@gmail.com.

Kellee Katagi (proofreading) has devoted most of her 20-plus-year writing and editing career to covering technology, fitness and nutrition, travel, and sports. A former managing editor of *SKI* magazine, she now smiths words from her Colorado home, where she lives with her husband and three kids. Email: kelkatagi@gmail.com.

Diana D'Abruzzo (proofreading) is a Virginia-based freelance editor with more than 20 years of experience in the journalism and book publishing industries. More information on her life and work can be found at dianadabruzzo.com.

Judy Le (symbol font and editorial assistance) is a magazine editor and school board member in Virginia, where she lives in a near-constant state of amazement with her husband and their son.

Julie Elman (design consulting) teaches courses in publication design, editorial illustration, and the creative process at a university in Ohio. She has more than 15 years' experience working as a visual journalist. Examples of her work can be found at julie-elman.com.

Jason Heuer (cover design) is an art director, designer, and artist living in New York City. He holds a BFA in Advertising & Design from the School of Visual Arts and was associate art director and recruiter at Simon & Schuster from 2004 to 2015. Currently he is principal at Jason Heuer LLC, a design and art studio in Rockaway. Find his work at jasonheuer.com.

Acknowledgments

Creating this book—and I mean *creating* it, from scratch—concept, philosophy, style, graphics, layout, design, covers—was a thrill. It began with my agent, Jim Levine, and the team at Simon & Schuster—Richard Rhorer, Priscilla Painton, and Jonathan Karp—who offered their full support for launching a new tech-book series in 2020. They said from the start they believed in the idea, and, wow, did they come through.

I was delighted that Jason Heuer (who designed the cover of my book *How to Prepare for Climate Change*) was willing to try his hand at a tech book; that Judy Le could assist with the index and draw symbols for our custom font; and that Diana D'Abruzzo, Julie Elman, and Kellee Katagi contributed their design and editing talents.

My most active collaborator, though, was Julie Van Keuren, who was responsible for basically every aspect of the book except for the writing and the figures. She came up with the title; she did the editing; she oversaw the design and proofing; she did the layout; she created the index; she chased down all the cross-references; she designed the custom symbol font; and she was my partner in creation on a thousand decisions, large and small.

Writing a book on a tight deadline always means withdrawing, to some degree, from the rest of the world. I'm grateful to my kids, Kell, Tia, and Jeffrey, for their tolerance during this period, and above all, to my bride, Nicki. Without her support, ideas, and belief in me, you'd be using Google searches right now to learn about your Mac. —*David Pogue*

Index

⊞ key. *See* **Mission Control**
⊞⊞ key. *See* **Launchpad**
 menu 3
⌘ (Command key) 25–26, 84
⌃ (Control key) 84
⋯ in menus 3–4
⌦ key 26
⌁, ⌁ keys 24
⌥ (Option key) 84
◀◀, ▶❚❚, ▶▶ playback keys 24
☼, ☼ keys 23
⬆ button. *See* **sharing**
⇧ (Shift key) 84
◀, ◀), ◀)) keys 24
4K videos 213

A

accents 141
accessibility 397–402
  Apple accessibility website 400
  dictation and Voice Control 379–385
  help with hearing 399
  help with mobility 399–400
  help with seeing 398–399
accounts 315–325
  account types 317–319
  choosing a picture 320–322
  creating 316–322
  defined 315–316
  deleting 322–323
  editing 322
  guest user account 321
  logging in 316
  sharing across accounts 324
  sharing with Drop Box 324
Activity Monitor 311–312
ads
  ad tracking in Safari 215–216
AirDrop 333–336
  controlling who sends you files 335
AirPlay 387
AirPods
  defined 12
  seamless switching 33
aliases 115–116
animated GIFs 225–226
AOL
  importing calendars 249
  setting up Mail 165–166
appearance
  dark and light modes 61–62
  new look of Big Sur 7–8
  screen saver 60
  wallpaper 58–60
Apple ID 150
Apple Music 240
Apple TV 387
Apple Watch login 153
AppleCare 370
Applications folder 110

F

G

H

I